Data Science for Beginners with MATLAB

A Practical, Hands-On Introduction to Data Analysis, Modeling, and Computational Thinking

Peter I. Kattan

Petra Books
www.PetraBooks.com

Data Science for Beginners with MATLAB

Peter I. Kattan, PhD

Correspondence about this book may be sent to the auth at one of the following two email addresses:

pkattan@PetraBooks.com

info@PetraBooks.com

Data Science for Beginners with MATLAB, written by Peter Kattan.

ISBN: 979-8-2956-2182-6

In Loving Memory of My Parents

Preface

This is a beginner-friendly book on ***data science using MATLAB***. It is designed as a practical introduction to data-driven thinking, computational analysis, and analytical modeling for students, researchers, and professionals who are new to the field of data science. The book begins with a foundational review of MATLAB as a computational environment and gradually develops the core skills needed for working with data, building models, and extracting meaningful information from numerical datasets.

The early chapters focus on essential computational tools: numerical operations, variables, mathematical functions, and data structures. These form the technical foundation of data science workflows, enabling readers to manipulate data, perform transformations, and construct analytical processes. The later chapters introduce descriptive analytics, exploratory data analysis, and predictive modeling, providing a structured pathway from raw data handling to data interpretation and model-based reasoning.

Rather than presenting data science as an abstract or overly theoretical discipline, this book treats it as a practical computational skillset. Emphasis is placed on learning by doing: using MATLAB commands to explore data, analyze patterns, compute relationships, and build simple predictive models. The goal is to develop computational intuition alongside analytical understanding, allowing readers to see data science as a natural extension of numerical reasoning and scientific thinking.

This book intentionally avoids advanced probability theory, complex machine learning algorithms, and specialized toolboxes. Instead, it focuses on core data science foundations: data representation, data transformation, exploratory analysis, statistical interpretation, and regression-based modeling. These form the conceptual backbone of modern data science and provide a strong base for later study in machine learning, artificial intelligence, and advanced analytics.

Specialized MATLAB toolboxes for advanced data science, machine learning, and statistics are powerful and valuable, but they are beyond the scope of a beginner's introduction. Readers interested in these advanced tools are encouraged to explore specialized references and advanced texts after mastering the foundational concepts presented here.

This book is intended for students and researchers across a wide range of disciplines, including engineering, physical sciences, biological sciences, behavioral sciences, social sciences, economics, environmental studies, and health sciences—fields where data-driven analysis is now central to discovery, decision-making, and innovation. No prior background in data science is assumed.

The purpose of this book is not simply to teach software commands, but to cultivate a data science mindset: structured thinking, analytical reasoning, model-based interpretation, and computational problem-solving. By learning how to work with data using MATLAB, readers develop transferable skills that apply across platforms, programming languages, and analytical environments.

As in the original work, simplicity, clarity, and accessibility remain guiding principles. The material is intentionally structured for beginners, allowing readers to build confidence step by step while developing real analytical capabilities. This book serves as both an introduction to data science and a foundation for future learning in advanced analytics, machine learning, and intelligent systems.

I hope this book provides readers with a clear, practical, and empowering entry into the world of data science—one that transforms data from numbers on a screen into meaningful insight, understanding, and knowledge.

The book includes nine chapters and about 150 exercises at the ends of chapters for students to practice.

I would like to thank my family members for their help and continued support without which this book would not have been possible. In this edition, I am providing two email addresses for my readers to contact me - pkattan@PetraBooks.com and info@PetraBooks.com. Please feel free to send your comments and suggestions.

February 2026 Peter I. Kattan

Table of Contents

1. Introduction to Data Science with MATLAB

This introductory chapter presents MATLAB as a foundational computational platform for modern data science. Rather than treating MATLAB only as a programming environment, this chapter frames it as a practical tool for data exploration, numerical reasoning, and analytical thinking. The commands introduced here form the basic language of data-driven work, enabling learners to interact with datasets, perform computations, and begin building analytical workflows that will later support statistical analysis, modeling, and data interpretation.

In this introductory chapter a short MATLAB tutorial is provided. This tutorial describes the basic MATLAB commands needed. More details about these commands will follow in subsequent chapters. Note that there are numerous free MATLAB tutorials on the internet – check references [1-28]. Also, you may want to consult many of the excellent books on the subject – check references [29-47]. This chapter may be skipped on a first reading of the book.

In this tutorial it is assumed that you have started MATLAB on your computer system successfully and that you are now ready to type the commands at the MATLAB prompt (which is denoted by double arrows “>>”). For installing MATLAB on your computer system, check the web links provided at the end of the book.

Entering scalars and simple operations is easy as is shown in the examples below:

```
>> 2*3+5

ans =

    11
```

The order of the operations will be discussed in subsequent chapters.

```
>> cos(60*pi/180)

ans =

    0.5000
```

The argument for the `cos` command and the value of `pi` will be discussed in subsequent chapters.

We assign the value of 4 to the variable `x` as follows:

```
>> x = 4

x =

     4

>> 3/sqrt(2+x)
```

```
ans =

     1.2247
```

To suppress the output in MATLAB use a semicolon to end the command line as in the following examples. If the semicolon is not used then the output will be shown by MATLAB:

```
>> y = 30;
>> z = 8;
>> x = 2; y-z;
>> w = 4*y + 3*z

w =

   144
```

MATLAB is case-sensitive, i.e. variables with lowercase letters are different than variables with uppercase letters. Consider the following examples using the variables x and X:

```
>> x = 1

x =

     1

>> X = 2

X =

     2
```

```
>> x

x =

     1

>> X

X =

     2
```

Use the `help` command to obtain help on any particular MATLAB command. The following example demonstrates the use of `help` to obtain help on the `det` command which calculated the determinant of a matrix:

```
>> help det
 DET    Determinant.
    DET(X) is the determinant of the square matrix X.

    Use COND instead of DET to test for matrix singularity.

    See also cond.

    Overloaded functions or methods (ones with the same name in other directories)
       help sym/det.m

    Reference page in Help browser
```

```
doc det
```

The following examples show how to enter matrices and perform some simple matrix operations:

```
>> x = [1 2 3 ; 4 5 6 ; 7 8 9]

x =

     1     2     3
     4     5     6
     7     8     9

>> y = [1 ; 0 ; -4]

y =

     1
     0
    -4

>> w = x*y

w =

   -11
   -20
   -29
```

Let us now see how to use MATLAB to solve a system of simultaneous algebraic equations. Let us solve the following system of simultaneous algebraic equations:

$$\begin{bmatrix} 3 & 5 & -1 \\ 0 & 4 & 2 \\ -2 & 1 & 5 \end{bmatrix} \begin{Bmatrix} x_1 \\ x_2 \\ x_3 \end{Bmatrix} = \begin{Bmatrix} 2 \\ 1 \\ -4 \end{Bmatrix} \quad (1.1)$$

We will use Gaussian elimination to solve the above system of equations. This is performed in MATLAB by using the backslash operator "\" as follows:

```
>> A= [3 5 -1 ; 0 4 2 ; -2 1 5]

A =

     3     5    -1
     0     4     2
    -2     1     5

>> b = [2 ; 1 ; -4]

b =

     2
     1
    -4

>> x = A\b

x =

   -1.7692
    1.1154
   -1.7308
```

It is clear that the solution is x_1 = -1.7692, x_2 = 1.1154, and x_3 = -1.7308. Alternatively, one can use the inverse matrix of A to obtain the same solution directly as follows:

```
>> x = inv(A)*b

x =

   -1.7692
    1.1154
   -1.7308
```

It should be noted that using the inverse method usually takes longer than using Gaussian elimination especially for large systems of equations.

Consider now the following 4 x 4 matrix D:

```
>> D = [1 2 3 4 ; 2 4 6 8 ; 3 6 9
12 ; -5 -3 -1 0]

D =

     1     2     3     4
     2     4     6     8
     3     6     9    12
    -5    -3    -1     0
```

We can extract the sub-matrix in rows 2 to 4 and columns 1 to 3 as follows:

```
>> E = D(2:4, 1:3)

E =
```

```
    2       4       6
    3       6       9
   -5      -3      -1
```

We can extract the third column of D as follows:

```
>> F = D(1:4, 3)

F =

    3
    6
    9
   -1
```

We can extract the second row of D as follows:

```
>> G = D(2, 1:4)

G =

    2       4       6       8
```

We can also extract the element in row 3 and column 2 as follows:

```
>> H = D(3,2)

H =

    6
```

We will now show how to produce a two-dimensional plot using MATLAB. In order to plot a graph of the function $y = f(x)$, we use the MATLAB command `plot(x,y)` after we have adequately defined both vectors `x` and `y`. The following is a simple example to plot the function $y = x^2 + 3$ for a certain range:

```
>> x = [1 2 3 4 5 6 7 8 9 10 11]

x =

  Columns 1 through 10

       1         2         3         4         5
6       7         8         9       10

  Column 11

      11

>> y = x.^2 + 3

y =

  Columns 1 through 10

       4         7        12        19        28
39      52        67        84     103

  Column 11

     124
```

```
>> plot(x,y)
```

Figure 1.1 shows the plot obtained by MATLAB. It is usually shown in a separate window. In this figure no titles are given to the x- and y-axes. These titles may be easily added to the figure using the xlabel and ylabel commands. More details about these commands will be presented in the chapter on graphs.

In the calculation of the values of the vector y, notice the dot "." before the exponentiation symbol "^". This dot is used to denote that the operation following it be performed element by element. More details about this issue will be discussed in subsequent chapters.

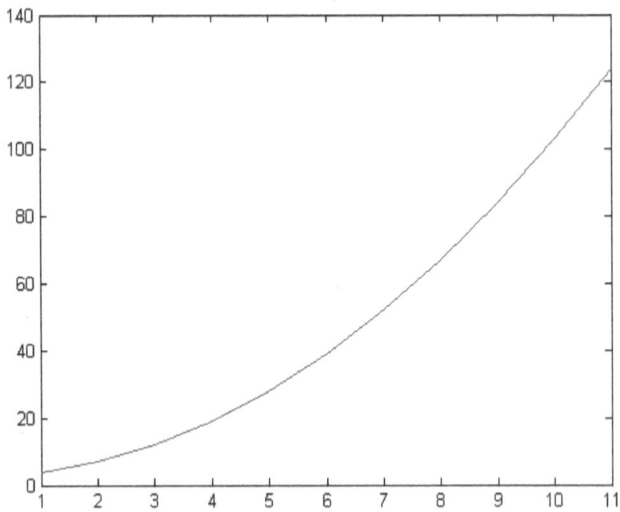

Figure 1.1: Using the MATLAB "plot" command

Finally, we will show how to make magic squares with MATLAB. A magic square is a square grid of numbers where the total of any row, column, or diagonal is the same. Magic squares are produced by MATLAB using the command `magic`. Here is a simple example to produce a 5 x 5 magic square:

```
>> magic(5)

ans =

    17    24     1     8    15
    23     5     7    14    16
     4     6    13    20    22
    10    12    19    21     3
    11    18    25     2     9
```

It is clear that the total of each row, column or diagonal in the above matrix is 65.

Exercises

Solve all the exercises using MATLAB. All the needed MATLAB commands for these exercises were presented in this chapter.

1. Perform the operation 3*4+6. The order of the operations will be discussed in subsequent chapters.

2. Perform the operation $\cos(5)$. The value of 5 is in radians.

3. Perform the operation $3\sqrt{6+x}$ for $x = 4$.

4. Assign the value of 5.2 to the variable `y`.

5. Assign the values of 3 and 4 to the variables `x` and `y`, respectively, then calculate the value of `z` where $z = 2x - 7y$.

6. Obtain help on the `inv` command.

7. Generate the following matrix: `A`

$$A = \begin{bmatrix} 1 & 0 & 2 & -3 \\ 0 & 5 & 2 & 2 \\ 1 & 2 & 3 & 4 \\ -2 & 0 & 1 & 3 \end{bmatrix}$$

8. Generate the following vector `b`

$$b = \begin{Bmatrix} 1 \\ 2 \\ 3 \\ 4 \end{Bmatrix}$$

9. Evaluate the vector `c` where $\{c\} = [A]\{b\}$ where `A` is the matrix given in Exercise 7 above and `b` is the vector given in Exercise 8 above.

10. Solve the following system of simultaneous algebraic equation using Gaussian elimination.

$$\begin{bmatrix} 5 & 2 \\ 1 & 3 \end{bmatrix} \begin{Bmatrix} x_1 \\ x_2 \end{Bmatrix} = \begin{Bmatrix} 3 \\ -1 \end{Bmatrix}$$

11. Solve the system of simultaneous algebraic equations of Exercise 10 above using matrix inversion.

12. Generate the following matrix X:

$$X = \begin{bmatrix} 1 & 0 & 6 \\ 1 & 2 & 3 \\ 4 & 5 & -2 \end{bmatrix}$$

13. Extract the sub-matrix in rows 2 to 3 and columns 1 to 2 of the matrix X in Exercise 12 above.

14. Extract the second column of the matrix X in Exercise 12 above.

15. Extract the first row of the matrix X in Exercise 12 above.

16. Extract the element in row 1 and column 3 of the matrix X in Exercise 12 above.

17. Generate the row vector x with integer values ranging from 1 to 9.

18. Plot the graph of the function $y = x^3 - 2$ for the range of the values of x in Exercise 17 above.

19. Generate a 4 x 4 magic square. What is the total of each row, column, and diagonal in this matrix.

2. Numerical Computation for Data Analysis

This chapter introduces numerical computation as the mathematical backbone of data science. Arithmetic operations in MATLAB are presented not simply as calculations, but as the building blocks of data transformation, feature construction, and quantitative analysis. These operations form the core of data preprocessing, algorithm development, and computational modeling workflows used throughout data science applications.

In this chapter we learn how to perform the simple arithmetic operations of addition, subtraction, multiplication, division, and exponentiation. The first four operations of addition, subtraction, multiplication, and division are performed in MATLAB using the usual symbols +, -, *, and /. The following are some examples:

```
>> 5+6

ans =

    11

>> 12-9

ans =
```

```
       3

>> 3*4

ans =

      12

>> 4/2

ans =

       2

>> 4/3

ans =

      1.3333
```

In the next chapter, we will show how to control the number of decimal digits that appear in the answer. The operation of exponentiation is performed using the symbol ^ as shown in the following examples:

```
>> 2^3

ans =

       8

>> 2^-3

ans =
```

```
      0.1250

>>
>> -2^3

ans =

      -8

>> -2^4

ans =

    -16

>> (-2)^4

ans =

      16
```

In the above examples, be careful where the minus sign is used. The constants π and ε are obtained in MATLAB as follows:

```
>> pi

ans =

      3.1416

>> eps

ans =
```

```
  2.2204e-016
```

In the above examples, the constant π is denoted by the command `pi` and represents the ratio of the perimeter of a circle to its diameter. The other constant ε is denoted by the command `eps` and represents the smallest number that MATLAB can handle.

In the above examples, only one arithmetic operation was performed in each line. Alternatively, we can perform multiple arithmetic operations in each line or command. The following are some examples:

```
>> 2+3-7

ans =

    -2

>> 3*4+5

ans =

    17

>> 2+5/4

ans =

    3.2500
```

In these cases, the order of operations is very important because the final computed result depends on this order. Precedence of computation is for the exponentiation operation, followed by multiplication and division, then followed by addition and subtraction. For example, in the examples above, multiplication is performed before addition and division is performed before addition. In case you need to perform addition before multiplication or division, the above two examples may be computed using parentheses as follows:

```
>> 3*(4+5)

ans =

    27

>> (2+5)/4

ans =

    1.7500
```

It is clear from the above examples that the operations inside the parentheses take precedence over any other operations. Several multiple operations may also be performed as follows:

```
>> (26+53)-(3^4 +5)*2/3

ans =

   21.6667
```

One should be careful when using parentheses. A change of the position of the parenthesis in the above example will change the computed results as follows:

```
>> (26+53)-(3^4 +5*2)/3

ans =

   48.6667
```

Arithmetic Operations with the MATLAB Symbolic Math Toolbox

Symbolic arithmetic operations may also be performed in MATLAB using the MATLAB Symbolic Math Toolbox. For the next examples, this toolbox needs to be installed with MATLAB. In order to perform symbolic operations in MATLAB, we need to define symbolic numbers[1] using the `sym` command. For example, ½ is defined as a symbolic number as follows:

```
>> sym(1/2)

ans =

1/2
```

In the above example, the number ½ is stored in MATLAB without any approximation using decimal digits like 0.5. Once used with the `sym` command, the computations will be performed algebraically or symbolically. For example, the

[1] Actually these are symbolic objects in MATLAB but objects are beyond the scope of this book.

following addition of two fractions is performed numerically as follows without the `sym` command:

```
>> 1/2+3/5

ans =

    1.1000
```

Adding parentheses will not affect the numerical computation in this case:

```
>> (1/2)+(3/5)

ans =

    1.1000
```

However, using the `sym` command of the MATLAB Symbolic Math Toolbox will result in the computation performed symbolically without the use of decimal digits;

```
>> sym((1/2)+(3/5))

ans =

11/10
```

Notice in the above example how the final answer was cast in the form of a fraction without decimal digits and without calculating a numerical value. In the addition of the two fractions above, MATLAB finds their common denominator and adds them by the usual procedure for rational numbers.

Notice also in the above four example outputs that symbolic answers are not indented but numerical answers are indented. In the next chapter, we will study variables and their use in MATLAB.

Exercises

Solve all the exercises using MATLAB. All the needed MATLAB commands for these exercises were presented in this chapter. Note that Exercises 19 and 21 require the use of the MATLAB Symbolic Math Toolbox.

1. Perform the addition operation 7+9.
2. Perform the subtraction operation 16-10.
3. Perform the multiplication operation 2*9.
4. Perform the division operation 12/3.
5. Perform the division operation 12/5.
6. Perform the exponentiation operation 3^5.
7. Perform the exponentiation operation 3*(-5).
8. Perform the exponentiation operation (-3)^5.
9. Perform the exponentiation operation -3^5.
10. Compute the value of $\frac{2\pi}{3}$.

11. Obtain the value of the smallest number that can be handled by MATLAB.

12. Perform the multiple operations 5+7-15.

13. Perform the multiple operations (6*7)+4.

14. Perform the multiple operations 6*(7+4).

15. Perform the multiple operations 4.5 + (15/2).

16. Perform the multiple operations (4.5 + 15)/2.

17. Perform the multiple operations (15 – 4 + 12)/5 – 2*(7^4)/100.

18. Perform the multiple operations (15 – 4) + 12/5 – (2*7)^4/100.

19. Define the number 2/3 as a symbolic number.

20. Perform the fraction addition (2/3) + (3/4) numerically.

21. Perform the fraction addition (2/3) + (3/4) symbolically.

3. Data Representation and Variable Management

This chapter introduces variables as fundamental data containers in data science. Variables in MATLAB represent data values, parameters, and intermediate results in analytical workflows. Understanding how data is stored, named, and manipulated provides the foundation for building structured data pipelines, reproducible analyses, and scalable computational models used in data-driven systems.

In this chapter we introduce variables and how to deal with them in MATLAB. If you do not use a variable, then MATLAB automatically stores the calculated result in the variable `ans`. The following is an example:

```
>> 3+2

ans =

     5
```

In the above example, the number 5 is stored automatically in the variable `ans`. This variable is short for "`answer`". If you wish to store the number 5 in another variable, then you need to specify it as follow:

```
>> x=3+2
```

```
x =

     5
```

In the above example, we used the variable x to store the result which is 5. You may define the variables x, y, z as follows without performing any operations:

```
>> x=1

x =

     1

>> y=-3

y =

    -3

>> z=3.52

z =

    3.5200
```

Note that MATLAB is case-sensitive with respect to variables. For example, the variables x and X are different. The following example illustrates this issue:

```
>> x=5

x =
```

```
     5

>> X=3

X =

     3

>> x

x =

     5

>> X

X =

     3
```

We may perform arithmetic operations on variables as shown in the following examples:

```
>> x=4

x =

     4

>> y=2.5

y =

    2.5000
```

```
>> z=x+y

z =

    6.5000

>> w=2*x-3*z+y

w =

    -9
```

Up till now, we have used single letters to denote variables. Any string of characters may be used to denote variables also. For example, we next use the variables `result` and `the_answer_is` as follows:

```
>> a=3

a =

     3

>> b=-2

b =

    -2

>> result = a*b

result =
```

```
    -6

>> the_answer_is = 2*a/b

the_answer_is =

    -3
```

Since MATLAB is case-sensitive, then the variables `width`, `WIDTH`, and `Width` are all different and store different results. You may add comments to your MATLAB commands and statements by using the symbol `%`. Comments are not executed and are ignored by MATLAB but are included for the benefit of the reader. For example, here is a comment:

```
>> % This is a comment
```

Comments are displayed in green color on the MATLAB screen. You may also add a comment on the same line as a command. The following is an example:

```
>> x = 6    % the value of x is
defined here

x =

     6
```

The following are the MATLAB rules for naming variables:

1. MATLAB is case-sensitive with respect to variables.
2. All variables must start with a letter.
3. You may have up to 31 characters in a variable.
4. Punctuation signs are not allowed in a variable.
5. The underscore sign "_" is allowed in a variable.
6. Both letters and digits are allowed in a variable.

For example, we may define variables `x1`, `x2`, and `x3` as follows:

```
>> x1 = 1

x1 =

     1

>> x2 = -2.7

x2 =

   -2.7000

>> x3 = 2*x1/x2

x3 =

   -0.7407
```

All variable used must be defined. For example, we get an error if we try to use a variable that is not defined. For example, the variable h is not defined as follows:

```
>> 3*h/2
??? Undefined function or variable
'h'.
```

The error is displayed in red color on the MATLAB screen. To find all the variables used in a MATLAB session, use the who command as follows:

```
>> who

Your variables are:

X              result             x1
z
a          the_answer_is    x2
ans        w                x3
b          x                y
```

The who command does not provide any details about the variables. The following is an example on using variables. If you know the cost and sale price of an item, calculate the profit. Here is the example:

```
>> cost = 100

cost =

   100

>> sale_price = 120
```

```
sale_price =

   120

>> profit = sale_price - cost

profit =

    20
```

Numbers in MATLAB are displayed with four decimal digits by default. For example:

```
>> 7/3

ans =

    2.3333
```

However, MATLAB performs the calculations internally using about 16 digits. You may ask MATLAB to display results with 16 digits using the command `format long` as follows:

```
>> format long
>> 7/3

ans =

   2.333333333333333

>> format short
>> 7/3

ans =
```

```
    2.3333
```

The command `format short` tells MATLAB to display four decimals again. Finally, you may ask MATLAB to show all variables used in a session with details regarding size, bytes[2], and class[3] using the `whos` command as follows:

```
>> whos
  Name        Size       Bytes  Class

  X                    1x1       8    double
array
  a                    1x1       8    double
array
  ans                  1x1       8    double
array
  b                    1x1       8    double
array
  cost                 1x1       8    double
array
  profit               1x1       8    double
array
  result               1x1       8    double
array
```

[2] The unit "bytes" is a measure of how much storage the variables take in the computer memory or disk. For example, a kilobyte (KB) is one thousand bytes and a megabyte (MB) is one million bytes.
[3] The "class" of a variable indicates the type of the variable. For example, all the variables in this session are of the class "double array". By default, all variables are stored as arrays (i.e. matrices) in MATLAB.

```
  sale_price       1x1      8   double
array
  the_answer_is   1x1      8   double
array
  w                1x1      8   double
array
  x                1x1      8   double
array
  x1               1x1      8   double
array
  x2               1x1      8   double
array
  x3               1x1      8   double
array
  y                1x1      8   double
array
  z                1x1      8   double
array

Grand total is 16 elements using
128 bytes
```

To clear all the variables in a session, use the `clear` command as follows:

```
>> clear
```

After you use the `clear` command, the command `who` or `whos` will not display any variables. Here is another example. If you know the radius of a circle, then calculate both the perimeter and area of the circle. The solution is:

```
radius =
```

```
   12.3000

>> perimeter = 2*pi*radius

perimeter =

   77.2832

>> area = pi*radius^2

area =

  475.2916
```

In the above example, we have used the constant `pi` that was defined in Chapter 2. Also, note the use of the two arithmetic operations of multiplication and exponentiation.

Variables with the MATLAB Symbolic Math Toolbox

You may also use symbolic variables[4] in MATLAB. For example, the variables `x` and `y` are defined as symbolic variables as follows:

```
>> x=1/2

x =

    0.5000
```

[4] Actually these are symbolic objects in MATLAB but objects are beyond the scope of this book.

```
>> y= 2/3

y =

    0.6667

>> sym(x)

ans =

1/2


>> sym(y)

ans =

2/3
```

Arithmetic operations can now be performed on these symbolic variables. Consider the following example where the variables x and y were defined as above :

```
>> z=sym(x-y)

z =

-1/6
```

In the above example, the variable z is used automatically as a symbolic variable. For symbolic operations, the important thing is to start each operation or calculation with the sym command.

Here is another example to calculate the volume of a sphere:

```
>> r=15/7

r =

    2.1429

>> sym(r)

ans =

15/7


>> volume = sym(4*pi*r^3/3)

volume =

4500*pi/343
```

In the above example, the final answer is given in the form of a fraction, $\frac{4500\pi}{343}$, with no decimal expressions and without calculating a numerical value (no units are used in this example). In case you need to have this value calculated numerically, then you can need to use the `double` command as follows:

```
>> double(volume)

ans =
```

```
41.2162
```

It is seen thus that the value of the volume of the sphere is 41.2162 (no units are used in this example). Finally, the variables `x` and `this` can be defined as symbolic variables without assigning a numerical value as follows:

```
>> x = sym('x')

x =

x


>> this = sym('this')

this =

this
```

In the next chapter, we will study mathematical functions and their use with variables.

Exercises

Solve all the exercises using MATLAB. All the needed MATLAB commands for these exercises were presented in this chapter. Note that Exercises 21-27 require the use of the MATLAB Symbolic Math Toolbox.

1. Perform the operation `2*3+7` and store the result in the variable `w`.

2. Define the three variables `a`, `b`, and `c` equal to `4`, `-10`, and `3.2`, respectively.

3. Define the two variables `y` and `Y` equal to `10` and `100`. Are the two variables identical?

4. Let `x = 5.5` and `y = -2.6`. Calculate the value of the variable `z = 2x-3y`.

5. In Exercise 4 above, calculate the value of the variable `w = 3y - z + x/y`.

6. Let `r = 6.3` and `s = 5.8`. Calculate the value of the variable `final` defined by `final = r + s - r*s`.

7. In Exercise 6 above, calculate the value of the variable `this_is_the_result` defined by `this_is_the_result = r^2 - s^2`.

8. Define the three variable `width`, `Width`, and `WIDTH` equal to 1.5, 2.0, and 4.5, respectively. Are these three variables identical?

9. Write the following comment in MATLAB: `This line will not be executed`.

10. Assign the value of `3.5` to the variable `s` then add a comment about this assignment on the same line.

11. Define the values of the variables `y1` and `y2` equal to 7 and 9 then perform the calculation `y3 = y1 - y2/3`. (Note: 2 in the formula is a subscript and should not be divided by 3).

12. Perform the operation `2*m - 5`. Do you get an error? Why?

13. Define the variables `cost` and `profit` equal to `175` and `25`, respectively, then calculate the variable `sale_price` defined by `sale_price = cost + profit`.

14. Define the variable `centigrade` equal to `28` then calculate the variable `fahrenheit` defined by `fahrenheit = (centigrade*9/5) + 32`.

15. Use the `format short` and `format long` commands to write the values of `14/9` to four decimals and sixteen digits, respectively.

16. Perform the `who` command to get a list of the variables stored in this session.

17. Perform the `whos` command to get a list of the variables stored in this session along with their details.

18. Clear all the variables stored in this session by using the `clear` command.

19. Calculate both the area and perimeter of a rectangle of sides 5 and 7. No units are used in this exercise.

20. Calculate both the area and perimeter of a circle of radius 6.45. No units are used in this exercise.

21. Define the symbolic variables x and z with values 4/5 and 14/17.

22. In Exercise 21 above, calculate symbolically the value of the symbolic variable y defined by $y = 2x - z$.

23. Calculate symbolically the area of a circle of radius 2/3 without obtaining a numerical value. No units are used in this exercise.

24. Calculate symbolically the volume of a sphere of radius 2/3 without obtaining a numerical value. No units are used in this exercise.

25. In Exercise 23 above, use the `double` command to obtain the numerical value of the answer.

26. In Exercise 24 above, use the `double` command to obtain the numerical value of the answer.

27. Define the symbolic variables `y` and `date` without assigning any numerical values to them.

4. Mathematical Functions for Data Modeling

This chapter presents mathematical functions as essential tools for data modeling and transformation. In data science, functions are used to extract patterns, model relationships, normalize data, and construct analytical features. MATLAB's built-in mathematical functions provide a computational framework for transforming raw data into structured, interpretable, and model-ready information.

In this chapter we introduce the use of mathematical functions in MATLAB. These functions include the square root function, the factorial, the trigonometric functions, the exponential functions, and the rounding and remainder functions[5]. We will start with the square root function[6] used with the MATLAB command `sqrt` as follows:

```
>> sqrt(2)

ans =

    1.4142
```

[5] MATLAB also has the special functions of mathematics like the Gamma function, the Error function, and Bessel functions but these are beyond the scope of this book.

[6] In this chapter we deal with square roots of positive numbers. Square roots of negative numbers are dealt with in the next chapter where we introduce complex numbers.

The above result gives the value of $\sqrt{2}$. The factorial of an integer is obtained using the MATLAB command `factorial` as follows:

```
>> factorial(5)

ans =

   120
```

All the trigonometric functions are also available within MATLAB. For example the sine and cosine functions are used with the MATLAB commands `sin` and `cos` as follows:

```
>> sin(30)

ans =

   -0.9880

>> sin(30*pi/180)

ans =

    0.5000

>> cos(30*pi/180)

ans =

    0.8660
```

Note that the argument for the trigonometric functions must be in radians. Check the first example above to see that the sine of the angle of 30 radians is -0.9880 but the sine of the angle of 30 degrees (converted into radians by multiplying it by $\frac{\pi}{180}$) is 0.5. The cosine of the angle of 30 degrees is seen above to be 0.8660. In addition, the tangent function is represented using the MATLAB command `tan` as follows where we compute the tangent of the angle of 30 degrees:

```
>> tan(30*pi/180)

ans =

    0.5774
```

The other trigonometric functions of secant, cosecant and cotangent are all represented in MATLAB using the commands `sec`, `csc`, and `cot`. In addition, the inverse trigonometric functions are represented by the MATLAB commands `asin`, `acos`, `atan`, `asec`, `acsc`, and `acot`. For example, the inverse tangent function is calculated as follows:

```
>> atan(1)

ans =

    0.7854
```

The angle returned in the above example is in radians. The hyperbolic trigonometric functions are represented by the MATLAB commands `sinh`, `cosh`, `tanh`, `sech`, `csch`, and `coth`. In addition, the inverse hyperbolic trigonometric functions are represented by the MATLAB commands `asinh`, `acosh`, `atanh`, `asech`, `acsch`, and `acoth`. Using these commands in MATLAB is straightforward. However, you need to consult a book on mathematics or trigonometry to check any limitations on the arguments of these functions. For example, if you try to find the tangent of the angle $\frac{\pi}{2}$ (90 degrees), then you will get a very large quantity as follows:

```
>> tan(pi/2)

ans =

  1.6331e+016
```

The reason for the strange result above is that the tangent function is not defined for the angle 90 degrees and the result given by MATLAB approaches infinity (thus we get a very large quantity).

The exponential and logarithmic functions are also represented by MATLAB. For example, the exponential function is represented by the MATLAB command `exp` as follows:

```
>> exp(1)

ans =

     2.7183
```

The above result gives the value of e^1. There are several logarithmic functions available in MATLAB. For example, the natural logarithm is represented by the command `log` and the logarithm to the base 10 is represented by the command `log10`. Here are two examples to illustrate this:

```
>> log(4)

ans =

    1.3863

>> log10(4)

ans =

    0.6021
```

The above two results give the values of $\ln 4$ and $\log_{10} 4$. We may also use the MATLAB command `log2` to obtain the logarithm to the base 2. For example, `log2(4)` will result in the value for $\log_2 4 = 2$.

Several rounding and remainder functions are also available in MATLAB. These functions are represented by the MATLAB commands `fix`, `floor`, `ceil`, `round`, `mod`, `rem`, and `sign`. For example, the rounding function `round` and the remainder function `rem` are used as follows:

```
>> round(4.57)
```

```
ans =

     5

>> rem(10,4)

ans =

     2
```

The above result gives the remainder when dividing 10 by 4. Finally, the absolute value function is represented by the MATLAB command `abs` as shown in the following example:

```
>> abs(-7)

ans =

     7
```

There are numerous other functions available in MATLAB like the complex functions which will be covered in the next chapter. Also, there are functions for coordinate transformations and there are number theoretic functions but these are beyond the scope of this book.

Mathematical functions can be combined using several commands on the same line in MATLAB and used with arithmetic operations. Here are some examples:

```
>> 2 + sqrt(5.3/3)
```

```
ans =

    3.3292

>> sin(cos(3.2))

ans =

   -0.8405

>> 2*log(2)/tan(pi/16)

ans =

    6.9694
```

The important point to consider when dealing with mathematical functions is to know what type of argument is valid for each function. For example, if you try to compute the natural logarithm of zero, then you will get an error or minus infinity because this function is not defined for zero. Here is the result:

```
>> log(0)
Warning: Log of zero.

ans =

  -Inf
```

Finally, mathematical functions can be used with variables. Here is an example:

```
>> a = 5
```

```
a =

     5

>> b = 2

b =

     2

>> c = b - sqrt(a+b/2)

c =

   -0.4495
```

Mathematical Functions with the MATLAB Symbolic Math Toolbox

Mathematical functions can also be used with the MATLAB Symbolic Math Toolbox. For example, the square root of 40 can be computed symbolically without performing any numerical calculations as follows:

```
>> sym(sqrt(40))

ans =

sqrt(40)


>> simplify(ans)

ans =
```

```
2*10^(1/2)
```

In the above, the `simplify` command is used to obtain the square root of 40 as $\sqrt{40} = 2\sqrt{10}$. To obtain a numerical result, use the `double` command applied on the previous result as follows:

```
>> double(ans)

ans =

    6.3246
```

The sine and cosine of the angle 30 degrees are computed symbolically as follows:

```
>> sym(sin(30*pi/180))

ans =

1/2


>> sym(cos(30*pi/180))

ans =

sqrt(3/4)
```

In addition, the tangent of the angle 30 degrees is obtained symbolically as follows:

```
>> sym(tan(30*pi/180))
```

```
ans =

sqrt(1/3)
```

Finally, the exponential and logarithm functions can also be used with Symbolic Math Toolbox. For example, the following command will compute $e^{2\pi}$ symbolically:

```
>> sym(exp(2*pi))

ans =

4710234414611992*2^(-43)
```

In order to obtain a numerical result, use the `double` command applied to the previous answer as follows:

```
>> double(ans)

ans =

  535.4917
```

In the next chapter, we will introduce complex numbers in MATLAB and how to handle them using complex functions

Exercises

Solve all the exercises using MATLAB. All the needed MATLAB commands for these exercises were presented in this chapter. Note that Exercises 19-25 require the use of the MATLAB Symbolic Math Toolbox.

1. Compute the square root of 10.

2. Compute the factorial of 7.

3. Compute the cosine of the angle 45 where 45 is in radians.

4. Compute the cosine of the angle 45 where 45 is in degrees.

5. Compute the sine of the angle of 45 where 45 is in degrees.

6. Compute the tangent of the angle 45 where 45 is in degrees.

7. Compute the inverse tangent of 1.5.

8. Compute the tangent of the angle $\frac{3\pi}{2}$. Do you get an error? Why?

9. Compute the value of exponential function e^3.

10. Compute the value of the natural logarithm $\ln 3.5$.

11. Compute the value of the logarithm $\log_{10} 3.5$.

12. Use the MATLAB rounding function `round` to round the value of 2.43.

13. Use the MATLAB remainder function `rem` to obtain the remainder when dividing 5 by 4.

14. Compute the absolute value of -3.6.

15. Compute the value of the expression $1.5-2\sqrt{6.7/5}$.

16. Compute the value of $\sin^2 \pi + \cos^2 \pi$.

17. Compute the value of $\log_{10} 0$. Do you get an error? Why?

18. Let $x = \dfrac{3\pi}{2}$ and $y = 2\pi$. Compute the value of the expression $2\sin x \cos y$.

19. Compute the value of $\sqrt{45}$ symbolically and simplify the result.

20. Compute the value of $\sqrt{45}$ numerically.

21. Compute the sine of the angle 45 (degrees) symbolically.

22. Compute the cosine of the angle 45 (degrees) symbolically.

23. Compute the tangent of the angle 45 (degrees) symbolically.

24. Compute the value of $e^{\pi/2}$ symbolically.

25. Compute the value of $e^{\pi/2}$ numerically.

5. Data Structures: Vectors as Data Containers

This chapter introduces vectors as fundamental data structures in data science. Vectors represent datasets, feature arrays, and measurement sequences in analytical workflows. Learning to manipulate vectors enables learners to perform batch processing, statistical computation, data transformation, and algorithmic operations that are central to data analysis and machine learning systems.

In this chapter we introduce vectors[7] and how to deal with them in MATLAB. Vectors are stored in MATLAB as one-dimensional arrays. For example, here is a vector stored in the variable x:

```
>> x = [ 1 3 0 5 -2 ]

x =

     1     3     0     5    -2
```

In the above example, the vector x contains five elements. It is noticed that the elements of a vector in MATLAB are separated by spaces and enclosed between brackets. To extract the second element of the vector x, we use the notation x(2) as follows:

```
>> x(2)
```

[7] We mean here vectors of numbers or scalars. There are also vectors of strings and characters in MATLAB but these are beyond the scope of this book.

```
ans =

     3
```

Here is another example where we extract the fifth element of the vector x:

```
>> x(5)

ans =

    -2
```

Here is a vector y defined implicitly with five elements as follows:

```
>> y = pi*x

y =

    3.1416    9.4248         0    15.7080
-6.2832
```

In the above example, the vector y is obtained from the vector x by multiplying each element of the vector x by π. To extract the fourth element of the vector y, we use the notation y(4) as follows:

```
>> y(4)

ans =

   15.7080
```

As we have seen above, we use the notation `y(n)` where `y` is the variable where the vector is stored and `n` is an integer ranging from 1 to the length of the vector. We have to be careful when we use this notation – `n` has to be an integer. If we try to use `y(1.2)` then we obtain an error message as follows (because `1.2` is not an integer):

```
>> y(1.2)
??? Subscript indices must either be real positive integers or logicals.
```

In order to extract the first three elements of the vector `y`, we use the notation `y(1:3)` as follows:

```
>> y(1:3)

ans =

    3.1416    9.4248         0
```

The length of a vector is the number of elements contained in the vector. In order to obtain the length of a vector, we use the MATLAB command `length` as follows:

```
>> length(y)

ans =

     5
```

In order to add the total sum of the values of the elements of a vector, we use the MATLAB command `sum` as follows:

```
>> sum(y)

ans =

   21.9911
```

In order to find the minimum value and the maximum value of the elements of a vector, we use the MATLAB commands `min` and `max`, respectively, as follows:

```
>> min(y)

ans =

   -6.2832

>> max(y)

ans =

   15.7080
```

In the above examples of vectors, we constructed the vectors by listing their elements either explicitly or implicitly. There are two other ways to construct vectors in MATLAB. For example, the vector `a` is constructed (using parentheses) as follows:

```
>> a = (0:0.2:1)
```

```
a =

          0        0.2000        0.4000
0.6000     0.8000     1.0000
```

In the above example, the vector `a` contains all the real numbers between 0 and 1 with an increment of 0.2. Another way to construct vectors in MATLAB is to use the `linspace`[8] command (using parentheses) as follows:

```
>> b = linspace(0,5,10)

b =

  Columns 1 through 6

          0        0.5556        1.1111
1.6667     2.2222     2.7778

  Columns 7 through 10

    3.3333        3.8889        4.4444
5.0000
```

In the above example, the vector `b` is constructed such that it contains 10 elements between 0 and 5 that are equally spaced[9]. Ignore the MATLAB output where it says "`Columns ? to ?`" – this

[8] There is also a `logspace` command in MATLAB but its use is beyond the scope of this book. It basically deals with generating logarithmically spaced vectors.

[9] Actually, to be precise, it is called linearly spaced.

notation will become clear to us when we study matrices in the next chapter.

Two vectors can be joined together to form a new vector. In the following example, the two vectors a and b are joined together to form the new vector c:

```
>> a = [ 1 3 5]

a =

     1     3     5

>> b = [ 7 9 ]

b =

     7     9

>> c = [a b]

c =

     1     3     5     7     9
```

Based on the above example, we can also form a new vector d as follows:

```
>> d = [a b 11]

d =

     1     3     5     7     9
11
```

Next, we will discuss the arithmetic operations that are allowed with vectors. The operations of addition and subtraction can be performed on two vectors of the same length. Here is an example:

```
>> x = [0.1 0.3 -2 5]

x =

    0.1000        0.3000       -2.0000
5.0000

>> y = [0.25 0.4 1 0]

y =

    0.2500        0.4000        1.0000
0

>> x+y

ans =

    0.3500        0.7000       -1.0000
5.0000

>> x-y

ans =

   -0.1500       -0.1000       -3.0000
5.0000
```

If you try to add or subtract two vectors of different lengths, then you will get an error message. Here is an example:

```
>> a = [1 5 3 5 6]

a =

     1     5     3     5     6

>> b = [-2 3 5 7]

b =

    -2     3     5     7

>> a+b
??? Error using ==> plus
Matrix dimensions must agree.

>> a-b
??? Error using ==> minus
Matrix dimensions must agree.
```

The operation of multiplication is a little more complicated for vectors. Even if you try to multiply two vectors of the same length, you will get an error message. Here is an example using the vectors `x` and `y` of the same length defined above:

```
>> x*y
??? Error using ==> mtimes
Inner matrix dimensions must agree.
```

Note that the operation of division is absolutely not defined for vectors – you cannot divide one vector by another. However, you can perform the new operations of element-by-element multiplication and element-by-element division of vectors in MATLAB using the vectors x and y defined above as follows:

```
>> x.*y

ans =

    0.0250         0.1200         -2.0000
0

>> x./y
Warning: Divide by zero.

ans =

    0.4000         0.7500         -2.0000
Inf
```

Note in the above example that the symbols for multiplication and division are preceded by a dot. The dot tells MATLAB to perform the operations element by element. Note that in order to carry these element-by-element operations, the two vectors must be of the same length.

There is another type of multiplication that is defined for vectors. It is called the scalar product, the inner product, or the dot product[10] – three different names for the same operation. This product is obtained in MATLAB for the two vectors x and y defined above as follows:

```
>> x*y'

ans =

   -1.8550
```

The final outcome of this operation is always a scalar or a number – not a vector. In addition, the two vectors that are used must be of the same length. Note the use of the prime on the second vector in the operation above. The use of the prime indicates the transpose operation which we will discussed in detail in the next chapter on matrices.

There are other simpler operations that can be performed on vectors. The operations of scalar addition and scalar subtraction can be performed as in the following example:

```
> s = [1 4 -3 2 2]

s =

     1     4    -3     2     2

>> s+4
```

[10] The reader is advised to consult a book on vector analysis for more details about this product.

```
ans =

     5     8     1     6     6

>> s-3

ans =

    -2     1    -6    -1    -1
```

In the above example, the number 4 is added to each element of the vector while the number 3 is subtracted from each element of the vector. In addition, the operations of scalar multiplication and scalar division can also be performed on the vector s defined above as follows:

```
>> s*2

ans =

     2     8    -6     4     4

>> s/5

ans =

    0.2000        0.8000        -0.6000
0.4000      0.4000
```

In the first example above, each element of the vector s is multiplied by the number 2 while in the second example, each element of the vector s is divided by the number 5. The above four operations can also be combined together using the vector s defined above as follows:

```
>> 2+ s/2.4

ans =

   2.4167    3.6667    0.7500    2.8333    2.8333
```

Mathematical functions can also be used with vectors. Here is an example where we use the three trigonometric functions on the vector a defined below:

```
>> a = [0 pi/2 pi 3*pi/2 2*pi]

a =

         0    1.5708    3.1416
4.7124    6.2832

>> sin(a)

ans =

         0    1.0000    0.0000    -
1.0000   -0.0000

>> cos(a)

ans =

    1.0000    0.0000   -1.0000    -
0.0000    1.0000
```

```
>> tan(a)

ans =

  1.0e+016 *

         0         1.6331         -0.0000
0.5444    -0.0000
```

In the above example, the trigonometric operations are carried out element-by-element. Here is an example with the same vector a where we use the exponential function:

```
>> exp(a)

ans =

    1.0000          4.8105          23.1407
111.3178   535.4917
```

In the above example, the exponential function is carried out element-by-element. Here is another example where we use the square root functions on the vector a defined above as follows:

```
>> sqrt(a)

ans =

         0         1.2533          1.7725
2.1708      2.5066
```

Note that the square root operation was performed element by element in the example above.

The last operation that we will discuss is the exponentiation operation. If we try to evaluate the quantity 2^a where a is the vector defined above, then we will receive an error message as follows:

```
>> 2^a
??? Error using ==> mpower
Matrix must be square.
```

In order to carry out the exponentiation operation element by element, we need to include the dot symbol before the exponentiation symbol as follows:

```
>> 2.^a

ans =

    1.0000        2.9707        8.8250
26.2162   77.8802
```

In the above example, the exponentiation operation is carried out element by element.

There are certain types of vectors that can be generated automatically by MATLAB. These types of vectors have specific commands for their generation. For example, the MATLAB commands `ones` and `zeros` produce vectors with elements of 1's and 0's, respectively. Here are two examples:

```
>> t = ones(1,7)

t =
```

```
       1          1          1          1          1
1      1

>> w = zeros(1,5)

w =

       0          0          0          0          0
```

The exact use of the above two commands will be discussed in detail in the next chapter on matrices. Elements of a vector can be sorted using the MATLAB command `sort` as follows:

```
>> x = [2 1 -3 5 3]

x =

       2          1         -3          5          3

>> sort(x)

ans =

      -3          1          2          3          5
```

Sorting in the above example has been obtained in an ascending order[11]. Finally, a vector with its element forming a random permutation can be generated using the `randperm` command as follows:

[11] Sorting can also be obtained in a descending order but this is beyond the scope of this book. This simple example of sorting is included to show the various capabilities of MATLAB.

```
>> randperm(7)

ans =

     2     7     4     3     6
5     1
```

In the above example, a random permutation of seven elements is obtained.

There are some MATLAB commands for statistics that can be used with vectors. For example, the MATLAB commands `range`, `mean`, and `median` can be used to obtain the statistical values of range, mean, and median for a set of numbers in a vector as follows:

```
>> w = [1 5 -2 0 4 6]

w =

     1     5    -2     0     4
6

>> range(w)

ans =

     8

>> mean(w)

ans =

    2.3333
```

```
>> median(w)

ans =

    2.5000
```

Vectors with the MATLAB Symbolic Math Toolbox

In this section, we will discuss the use of symbolic vectors in MATLAB. Symbolic vectors are vectors whose elements are handled algebraically without numerical computations. In the following example, we define a symbolic vector `a` of three elements using the MATLAB command `syms`:

```
>> syms x y z
>> a = [x y z]

a =

[ x, y, z]
```

Next, we carry out the following scalar addition operation on the vector `a` defined previously to obtain the new symbolic vector `b`:

```
>> b = 2+a

b =

[ x+2, y+2, z+2]
```

Next, we extract the second element of the vector b as follows:

```
>> b(2)

ans =

y+2
```

Next, we perform the addition of the two vectors a and b to get the new symbolic vector c as follows:

```
>> c = a+b

c =

[ 2*x+2, 2*y+2, 2*z+2]
```

Next, we perform the operation 2*a – 3*b/5 to obtain the new symbolic vector d as follows:

```
>> d = 2*a - 3*b/5

d =

[ 7/5*x-6/5, 7/5*y-6/5, 7/5*z-6/5]
```

Next, we obtain the dot product of the vector a and b as follows:

```
>> a*b'

ans =
```

```
x*(2+conj(x))+y*(2+conj(y))+z*(2+c
onj(z))
```

In the above example, notice that the result is a scalar quantity (a symbolic or algebraic expression) – not a vector. We can also perform the dot product in the following way by reversing the order of the two vectors:

```
>> b*a'

ans =

(x+2)*conj(x)+(y+2)*conj(y)+(z+2)*
conj(z)
```

Notice that the two dot products obtained above are different. They would be exactly the same if the variables `x`, `y`, and `z` were real variables. MATLAB assumes by default that these are complex variables thus obtaining two different results.

Next, we obtain the square root of the vector `a-b` element by element as follows:

```
>> sqrt(a-b)

ans =

[ i*2^(1/2), i*2^(1/2), i*2^(1/2)]
```

The symbolic vector obtained above is a complex vector, i.e. a vector with complex elements. Finally, we can obtain a numerical value for the above expression by using the MATLAB command double as follows:

```
>> double(ans)

ans =

   0 + 1.4142i   0 + 1.4142i   0 +
1.4142i
```

It is clear from the above result that the final vector is a complex vector with its elements being complex numbers. In the next chapter we will discuss the use of matrices in MATLAB.

Exercises

Solve all the exercises using MATLAB. All the needed MATLAB commands for these exercises were presented in this chapter. Note that Exercises 36-41 require the use of the MATLAB Symbolic Math Toolbox.

1. Store the vector [2 4 -6 0] in the variable `w`.

2. In Exercise 1 above, extract the second element of the vector `w`.

3. In Exercise 1 above, generate the vector `z` where $z = \frac{\pi}{2} w$.

4. In Exercise 3 above, extract the fourth element of the vector `z`.

5. In Exercise 3 above, extract the first three elements of the vector `z`.

6. In Exercise 3 above, find the length of the vector `z`.

7. In Exercise 3 above, find the total sum of the values of the elements of the vector `z`.

8. In Exercise 3 above, find the minimum and maximum values of the elements of the vector `z`.

9. Generate a vector `r` with real values between 1 and 10 with an increment of 2.5.

10. Generate a vector `s` with real values of ten numbers that are equally spaced between 1 and 100.

11. Form a new vector by joining the two vectors [9 3 -2 5 0] and [1 2 -4].

12. Form a new vector by joining the vector [9 3 -2 5 0] with the number 4.

13. Add the two vectors [0.2 1.3 -3.5] and [0.5 -2.5 1.0].

14. Subtract the two vectors in Exercise 13 above.

15. Try to multiply the two vectors in Exercise 13 above. Do you get an error message? Why?

16. Multiply the two elements in Exercise 13 above element by element.

17. Divide the two elements in Exercise 13 above element by element.

18. Find the dot product of the two vectors in Exercise 13 above.

19. Try to add the two vectors [1 3 5] and [3 6]. Do you get an error message? Why?

20. Try to subtract the two vectors in Exercise 20 above. Do you get an error message? Why?

21. Let the vector w be defined by `w` = [0.1 1.3 -2.4]. Perform the operation of scalar addition `5+w`.

22. In Exercise 22 above, perform the operation of scalar subtraction `-2-w`.

23. In Exercise 22 above, perform the operation of scalar multiplication `1.5*w`.

24. In Exercise 22 above, perform the operation of scalar division `w/10`.

25. In Exercise 22 above, perform the operation `3 - 2*w/5`.

26. Define the vector `b` by `b` = [0 pi/3 2pi/3 pi]. Evaluate the three vectors $\sin b$, $\cos b$, and $\tan b$ (element by element).

27. In Exercise 26 above, evaluate the vector e^b (element by element).

28. In Exercise 26 above, evaluate the vector $\sqrt{b}$ (element by element).

29. Try to evaluate the vector 3^b. Do you get an error message? Why?

30. Perform the operation in Exercise 29 above element by element?

31. Generate a vector of 1's with a length of 4 elements.

32. Generate a vector of 0's with a length of 6 elements.

33. Sort the elements of the vector [0.35 -1.0 0.24 1.30 -0.03] in ascending order.

34. Generate a random permutation vector with 5 elements.

35. For the vector [2 4 -3 0 1 5 7], determine the range, mean, and median.

36. Define the symbolic vector x = [r s t u v].

37. In Exercise 36 above, perform the addition operation of the two vectors x and [1 0 -2 3 5] to obtain the new symbolic vector y.

38. In Exercise 37 above, extract the third element of the vector `y`.

39. In Exercise 37 above, perform the operation `2*x/7 + 3*y`.

40. In Exercise 37 above, perform the dot products `x*y'` and `y*x'`. Are the two results the same? Why?

41. In Exercise 37 above, find the square root of the symbolic vector `x+y`.

6. Descriptive Analytics and Exploratory Data Analysis (EDA)

This chapter introduces descriptive analytics as a core component of data science. Measures of central tendency and dispersion are presented as tools for understanding data distributions, identifying patterns, and characterizing datasets. These concepts form the foundation of exploratory data analysis (EDA), which is essential for data interpretation, quality assessment, and model preparation.

In this chapter the basic numerical descriptive measures of statistical data are discussed. Measures of central tendency and measures of dispersion are presented including the mean, median, variance, and standard deviation. All the respective MATLAB commands[12] used for statistics in this chapter are illustrated with examples. There is no coverage of probability theory – only basic statistical concepts.

There are three measures of central tendency that are used most often. These are the arithmetic mean, median, and mode[13]. Others are also discussed like the geometric mean and the harmonic mean. As for the measures of dispersion, the two most used

---

[12] All the MATLAB commands used for statistics in this chapter are found in the main MATLAB package. No resort is made to the Statistics Toolbox that is purchased separately. The Statistics Toolbox includes specialized and advanced commands for Statistics but these are beyond the scope of this book.

[13] The three measures: range, mean, and median were briefly described at the end of Chapter 6 on Vectors.

measures are the variance and standard deviation. The percentiles and quartiles are also used but these two do not measure dispersion.

Let us start by listing the data for the example to be used throughout this chapter. The heights of 30 people in a neighborhood are measured in centimeters and are listed below:

118	147	158	162	111
150	162	155	143	170
192	181	151	144	137
171	180	190	163	165
157	148	188	182	170
170	181	179	182	160

The above data are stored in a vector in MATLAB. Let us call this vector `x`. The following is the relevant MATLAB command:

```
>> x = [118 147 158 162 111 150 162 155
143 170 192 181 151 144 137 171 180 190
163 165 157 148 188 182 170 170 181 179
182 160]

x =

  Columns 1 through 12

   118   147   158   162   111   150
162   155   143   170   192   181

  Columns 13 through 24

   151   144   137   171   180   190
163   165   157   148   188   182

  Columns 25 through 30
```

```
   170    170    181    179    182    160
```

Alternatively, and if the data set is too long, one my store it in a text file (let us call it file_name), and then import the text file into MATLAB using the command `load` as follows:

```
x = load(file_name)
```

Let us first determine the size of the data set, i.e. the number of data points stored in the vector `x`. For this purpose, we use the MATLAB command `size` as follows:

```
>> size(x)

ans =

     1    30
```

Thus it is clear that the vector X has 30 data points. Next, let us sort the data in the vector X using the MATLAB command `sort` as follows:

```
>> sort(X)

ans =

  Columns 1 through 10

   111    118    137    143    144    147
148    150    151    155

  Columns 11 through 20

   157    158    160    162    162    163
165    170    170    170
```

```
  Columns 21 through 30

   171    179    180    181    181    182
182    188    190    192
```

The measures of central tendency all try to find a value from the data set that lies at the center of the data. Different ways are used to do that involving the arithmetic mean, geometric mean, harmonic mean, median, mode, etc.

The arithmetic mean[14] of the heights is obtained by using the MATLAB command `mean` as follows:

```
>> mean(x)

ans =

  162.2333
```

The mean of a data set consisting of a number of data points is the arithmetic average of the data set. The mean is usually denoted by the symbol $\overline{x}$ for a data set consisting of n points x_i , $i = 1,2,3,.....,n$. The mean is usually calculated by hand using the following formula (however the use of the simple formula below is left to the reader; is suffices to use the MATLAB command as shown above):

[14] Two other types of "mean" are also available in MATLAB. The geometric mean and the harmonic mean are calculated using the MATLAB commands geomean and harmean, respectively, but these are beyond the scope of this book.

$$\bar{x} = \frac{\sum_{i=1}^{n} x_i}{n}$$

The median of a data set consisting of a number of data points is literally the middle value when the data points are sorted or ordered. In this case, half the values are above the median and the other half is below the median. The median of the heights is obtained by using the MATLAB command `median` as follows:

```
>> median(x)

ans =

  162.5000
```

The mode of the heights is obtained by using the MATLAB command `mode` as follows:

```
>> mode(x)

ans =

   170
```

It is noted that if the data is symmetrical, i.e. if it is symmetrically distributed, then the values of the arithmetic mean, the median, and the mode will be identical.

The measures of dispersion are values not from the data set, but are calculated through different formulas that try to show how the data differ or deviate from the central value. Such measures include the average deviation, variance, standard deviation, etc.

The range is simply the difference between the largest data value and smallest data value. The range of the give data is obtained using MATLAB using the command `range` as follows:

```
>> range(x)

ans =

    81
```

The interquartile range is the difference between the third quartile value and the first quartile value. A quartile (or a quarter) is one fourth of the data. Thus, the first quartile is the data value that has 25% of the data below it, while the third quartile is the data value that has 75$ of the data below it. Actually, the interquartile range is a measure of the middle 50% of the data. The interquartile range (IQR) of the give data is obtained using MATLAB using the command `iqr` as follows:

```
>> iqr(x)

ans =

    30
```

The variance s^2 of the data is a measure of the spread of the data. The variance is calculated by hand using the following formula:

$$s^2 = \frac{\sum_{i=1}^{n} (x_i - \bar{x})^2}{n-1}$$

The variance of the data is obtained by using the MATLAB command `var` as follows:

```
>> var(x)

ans =

  393.2195
```

The standard deviation s is also a measure of the spread of the data but is much more useful than the variance. The standard deviation is calculated manually using the following formula:

$$s^2 = \sqrt{\frac{\sum_{i=1}^{n} (x_i - \bar{x})^2}{n-1}}$$

The standard deviation of the data is obtained by using the MATLAB command `std` as follows:

```
>> std(x)

ans =

   19.8298
```

In all the above calculation, it is assumed that the data set of heights is a sample. There is no mention of the population, but it is noted that there are variations of the above measures that apply to the population, instead of the sample. Such measures would include the population mean and the population standard deviation, but these are beyond the scope of this work.

Next, we disucss measures of position, basically the quartiles, the percentiles, and the standardized scores. Quartiles divide the data into four equal parts, each of which contains 25% of the data. A percentile is one of 99 values that divide a set of data such that it gives the percentage of data values that lie below the given value. The `pth` percentile P is defined as the value of the data such that p percent of the data are less than P and (100-p) percent of the data are greater than P. The three most popular `pth` percentiles are the 25`th` percentile (called the first quartile), the 50`th` percentile (called the second or middle quartile), and the 75`th` percentile (called the third quartile).

The following are some examples using MATLAB showing how use the different commands to calcualate some percentiles and quartiles for the data set used in this chapter:

The 0.30 quantile is calculated using MATLAB using the command `quantile` as follows:

```
>> quantile(x,0.30)

ans =

   153
```

The 0.50 quantile is calculated using MATLAB using the command `quantile` as follows:

```
>> quantile(x,0.50)

ans =

  162.5000
```

The 0.70 quantile is calculated using MATLAB using the command `quantile` as follows:

```
>> quantile(x,0.70)

ans =

   175
```

The 42nd percentile is calculated using MATLAB using the command `prctile` as follows:

```
>> prctile(x,42)

ans =

  160.2000
```

The 65th percentile is calculated using MATLAB using the command `prctile` as follows:

```
>> prctile(x,65)

ans =

   170
```

The 87th percentile is calculated using MATLAB using the command `prctile` as follows:

```
>> prctile(x,87)

ans =

   182
```

Finally, standardized scores (zscores) are used as a measue to to give the position of a data value relative to its mean and standard deviation. The zscore z is calculated by hand using the following formula:

$$z = \frac{x - \overline{x}}{s}$$

Based on the above formula, it is noted that for each data value x, a corresponding zscore value z is calculated. Thus, for a data set consisting of n data points, one can calculate n different zscores. The manual use of the above formula is left to the reader but calculating zscores using MATLAB will be illustrated below.

Standardized scores (also called zscores) can be calculated using MATLAB using the command `zscore` as follows:

```
>> zscore(x)

ans =

  Columns 1 through 7

   -2.2307   -0.7682   -0.2135   -0.0118
-2.5837   -0.6169   -0.0118

  Columns 8 through 14
```

```
   -0.3648    -0.9699     0.3917     1.5011
0.9464    -0.5665    -0.9195

  Columns 15 through 21

   -1.2725     0.4421     0.8960     1.4003
0.0387     0.1395    -0.2639

  Columns 22 through 28

   -0.7178     1.2994     0.9968     0.3917
0.3917     0.9464     0.8455

  Columns 29 through 30

    .9968     -0.1126
```

Other descriptive statistical measures used are the maximum value and minimum value of the data set. These three measures are illustrated below by using the MATLAB commands `max` and `min` as follows:

```
>> max(x)

ans =

   192

>> min(x)

ans =

   111
```

Finally, additional statistical measures can be calculated such as skewness, kurtosis, and moment using the MATLAB commands `skewness`, `kurtosis`, and `moment`, respectively. But these commands will not be illustrated here as they are beyond the scope of this beginner's book.

In the remainder of this chapter, we discuss how to calculate Pearson's correlation coefficient[15]. This linear correlation coefficient is used to meaure the strength of a linear relationship between two variables. This measurement has limits at -1 and +1. The endpoints indicate a perfect correlation. The farther the measurement is from zero, the stronger the relationship between the two variables.

The Pearson correlation coefficient is illustrated in the next example. For this purpose, consider two sets of data or observations that need to be compared. Each set of data describes an independent variable. The value of the correlation coefficient lies between +1 and -1 inclusive. Total positive correlation exists when the value of the correlation coefficient is equal to +1. No correlation exists when the value of the correlation coefficient is equal to 0. Total negative correlation exists when the value of the correlation coefficient is equal to -1. If the value of the correlation coefficient is positive, the positive correlation exists between the two variables. On the other hand, negative correlation exists between the two variables if the value of the correlation coefficient is negative. The two data sets for this example are shown below.

Set A Data:

---

[15] Other types of correlation coefficients exist in the statistics literature, but these are beyond the scope of this book.

11 14 17 20 24 56
60
66 70 72 77 81 90
100
101 110 116 118 120 122
140

Set B Data:

1.4 2.5 2.6 3.2 3.3 3.5
4.1
4.5 4.7 5.0 5.6 6.0 6.6
7.0
7.8 8.0 9.0 9.9 10.1 11.1
12.0

In order to conduct a correlation analysis between the two sets of data A and B, use the MATLAB command `corrcoef` to calculate the Pearson correlation coefficient of these two sets of data as follows:

```
>> x = [11 14 17 20 24 56 60 66 70 72 77
81 90 100 101 110 116 118 120 122 140]

x =

  Columns 1 through 12

    11      14      17      20      24      56
60     66     70     72     77     81

  Columns 13 through 21

    90     100     101     110     116     118
120    122    140
```

```
>> y = [1.4 2.5 2.6 3.2 3.3 3.5 4.1 4.5
4.7 5.0 5.6 6.0 6.6 7.0 7.8 8.0 9.0 9.9
10.1 11.1 12.0]

y =

  Columns 1 through 7

    1.4000      2.5000      2.6000      3.2000
3.3000     3.5000     4.1000

  Columns 8 through 14

    4.5000      4.7000      5.0000      5.6000
6.0000     6.6000     7.0000

  Columns 15 through 21

    7.8000      8.0000      9.0000      9.9000
10.1000    11.1000    12.0000

>> corrcoef(x,y)

ans =

    1.0000    0.9576
    0.9576    1.0000
```

Thus, the value of the correlation coefficient in this example is 0.9576 indicating a strong positive correlation. Alternatively, instead of using the MATLAB command `corrcoef` directly, one can use the formula for Pearson's correlation coefficient, and calculate the value of the coefficient by hand. The formula is given by the following equation for two sets x_i and y_i of n data points:

$$r = \frac{n\sum_{i=1}^{n} x_i y_i - \sum_{i=1}^{n} x_i \sum_{i=1}^{n} y_i}{\sqrt{n\sum_{i=1}^{n} x_i^2 - \left(\sum_{i=1}^{n} x_i\right)^2}\sqrt{n\sum_{i=1}^{n} y_i^2 - \left(\sum_{i=1}^{n} y_i\right)^2}}$$

As an example of using MATLAB, we will illustrate how to use the above formula in order to calculate the value of the correlation coefficient a second time. In order to use MATLAB to compute the above formula for the two data sets in the example above, one needs first to calculate the sums $\sum_{i=1}^{n} x_i$, $\sum_{i=1}^{n} y_i$, $\sum_{i=1}^{n} x_i^2$, $\sum_{i=1}^{n} y_i^2$, and $\sum_{i=1}^{n} x_i y_i$. For this purpose, Table 12.1 shows each sum along with the required MATLAB command associated with it:

Table 12.1: MATLAB Command Required for Each Respective Sum

Sum	MATLAB Command
x_i	`x`
y_i	`y`
$\sum_{i=1}^{n} x_i$	`sum(x)`
$\sum_{i=1}^{n} y_i$	`sum(y)`
$\sum_{i=1}^{n} x_i^2$	`mxx=x.*x` `sum(sxx)`

$\sum_{i=1}^{n} y_i^2$	myy=y.*y sum(syy)
$\sum_{i=1}^{n} x_i y_i$	mxy=x.*y sum(sxy)

The following are the MATLAB commands needed to calculate Pearson's correlation coefficient for the example above using the formula given above:

Step One: Enter the data points given in the Example:

```
>> x = [11 14 17 20 24 56 60 66 70 72 77
81 90 100 101 110 116 118 120 122 140]

x =

  Columns 1 through 12

    11      14      17      20      24      56
60    66    70    72    77    81

  Columns 13 through 21

    90     100     101     110     116     118
120   122   140

>> y = [1.4 2.5 2.6 3.2 3.3 3.5 4.1 4.5
4.7 5.0 5.6 6.0 6.6 7.0 7.8 8.0 9.0 9.9
10.1 11.1 12.0]

y =

  Columns 1 through 7

    1.4000      2.5000      2.6000      3.2000
3.3000     3.5000     4.1000
```

```
  Columns 8 through 14

    4.5000       4.7000       5.0000       5.6000
6.0000     6.6000     7.0000

  Columns 15 through 21

    7.8000       8.0000       9.0000       9.9000
10.1000    11.1000    12.0000
```

<u>Step Two:</u> Calculate the five different sums needed to compute the coefficient using the formula:

2.1 Calcuate the sum $\sum_{i=1}^{n} x_i$:

```
>> sumx = sum(x)

sumx =

        1585
```

2.2 Calculate the sum $\sum_{i=1}^{n} y_i$:

```
>> sumy = sum(y)

sumy =

  127.9000
```

2.3: Calculate the sum $\sum_{i=1}^{n} x_i^2$:

```
>> mxx = x.*x

mxx =
```

```
  Columns 1 through 5

         121              196              289
400          576

  Columns 6 through 10

        3136             3600             4356
4900         5184

  Columns 11 through 15

        5929             6561             8100
10000        10201

  Columns 16 through 20

       12100            13456            13924
14400        14884

  Column 21

       19600

>> sumxx = sum(mxx)

sumxx =

      151913
```

2.4: Calculate the sum $\sum_{i=1}^{n} y_i^2$:

```
>> myy = y.*y

myy =

  Columns 1 through 7
```

```
   1.9600     6.2500     6.7600    10.2400
10.8900   12.2500   16.8100

  Columns 8 through 14

   20.2500    22.0900    25.0000    31.3600
36.0000   43.5600   49.0000

  Columns 15 through 21

   60.8400    64.0000    81.0000    98.0100
102.0100  123.2100  144.0000

>> sumyy = sum(myy)

sumyy =

  965.4900
```

2.5: Calculate the sum $\sum_{i=1}^{n} x_i y_i$:

```
>> mxy = x.*y

mxy =

   1.0e+03 *

  Columns 1 through 7

    0.0154     0.0350     0.0442     0.0640
0.0792     0.1960     0.2460

  Columns 8 through 14

    0.2970     0.3290     0.3600     0.4312
0.4860     0.5940     0.7000

  Columns 15 through 21
```

```
    0.7878      0.8800      1.0440      1.1682
1.2120     1.3542     1.6800

>> sumxy = sum(mxy)

sumxy =

   1.2003e+04
```

2.6: Calculate the square $\left(\sum_{i=1}^{n} x_i\right)^2$:

```
>> sumx2 = sumx*sumx

sumx2 =

     2512225
```

2.7: Calculate the square $\left(\sum_{i=1}^{n} y_i\right)^2$:

```
>> sumy2 = sumy*sumy

sumy2 =

   1.6358e+04
```

2.8: Calculate the number of data points in the data set using the `size` command:

```
>> n = size(x)

n =

     1    21

>> n1 = n(2)
```

```
n1 =

    21
```

Step Three: Use the formula to calculate the value of the correlation coefficient by substituting the values of the sums obtained above:

```
>>   r   =   (n1*sumxy   -   sumx*sumy)   /
(sqrt(n1*sumxx-sumx2)*sqrt(n1*sumyy-
sumy2))

r =

    0.9576
```

It is clear from the result above that the correlation between the two data sets is positive and high. The value of the coefficient is very close to 1.0. It is noted that the exact same value of the correlation coefficient (i.e. 0.9576) is obtained also using the MATLAB command corrcoef as was shown previously (at the beginning of this example).

Exercises

Solve Exercies 1 - 8 using the following set of 25 data points using MATLAB:

25	18	17	24	11
10	20	30	25	12
11	12	15	28	33
20	12	20	22	30
15	18	22	19	31

1. Store the data set in a vector called x using MATLAB.

2. Sort the data set using MATLAB and find its size (i.e. the number of data points) using MATLAB.

3. Calculate the mean, median and mode for the data set using MATLAB.

4. Calculate the range, the interquantile range, the variance, and the standard deviation for the data set using MATLAB.

5. Calculate the maximum and minimum values using MATLAB.

6. Calculate the 0.30 quantile, the 0.50 quantile, and the 0.70 quantile for this data set.

7. Calculate the 47nd percentile, the 55th percentile, and the 80th percentile for this data set.

8. Calculate the standardized scores (zscore) for this data set.

Solve Exercies 9-11 using the following two sets of data points using MATLAB:

Data Set One:

1.24 2.56 3.45 4.66 3.56 4.90

3.33 2.45 2.34 2.00 1.45 1.56

Data Set Two:

600 450 320 300 430 330
470 570 340 200 250 390

9. Calculate Pearson's correlation coefficient for the two sets of data points given above directly in MATLAB using the command `corrcoef`.

10. Calculate Pearson's correlation coefficient for the aboe two sets of data points using MATLAB but indirectly by using the formula given in this chapter. Note that you need to calculate the various sums appearing in the formula first.

11. Compare the two results that you obtained for Pearson's correlation coefficient from Exercises # 9 and # 10 above.

7. Statistical Data Exploration and Interpretation

This chapter continues the development of statistical thinking in data science by focusing on deeper data exploration techniques. Statistical measures and visual analysis tools are framed as methods for understanding variability, structure, and trends in data. These tools support data-driven decision-making and model development across scientific and applied domains.

In this chapter the basic methods of graphical visualization of statistical data are discussed. Several types of graphs designed for statistical data are presented such as histograms, frequency diagrams, scatter diagram, and pie charts. All the respective MATLAB commands[16] used for statistics in this chapter are illustrated with examples.

In order to demonstrate the various graphical measures used in statistics, let us start by again listing the data for the same example of Chapter 12 to be used throughout this chapter. The heights of 30 people in a neighborhood are measured in centimeters and are listed below:

[16] All the MATLAB commands used for statistics in this chapter are found in the main MATLAB package and no resort is made to the Statistics Toolbox that is purchased separately. The Statistics Toolbox includes more specialized and advanced commands for Statistics but these are beyond the scope of this book.

118	147	158	162	111
150	162	155	143	170
192	181	151	144	137
171	180	190	163	165
157	148	188	182	170
170	181	179	182	160

The above data are stored in a vector in MATLAB. Let us call this vector `x`. The following is the relevant MATLAB command (repeated from Chapter 12):

```
>> x = [118 147 158 162 111 150 162 155
143 170 192 181 151 144 137 171 180 190
163 165 157 148 188 182 170 170 181 179
182 160]

x =

  Columns 1 through 12

   118     147     158     162     111     150
162    155    143    170    192    181

  Columns 13 through 24

   151     144     137     171     180     190
163    165    157    148    188    182

  Columns 25 through 30

   170    170    181    179    182    160
```

The bar chart (also called bar graph) of the data is obtained using the MATLAB command `bar` as follows. The resulting graph is shown in Figure 13.1. The bar chart shows the counts for each category of heights. Each category of information is depicted in a separate bar. Individual bars are separated by a space. The height of each bar represent the count or number of heights for that specific category.

```
>> bar(x)
>> hold on
>> xlabel('heights')
>> ylabel('Number of People')
>> xlabel('Heights')
```

The histogram of the data is obtained using the MATLAB command `hist` as follows. The resulting graph is shown in Figure 13.2. Like the bar chart, the histogram plots data in bars, where the height of the bar represents the count or number of heights. There are no spaces between bars in a histogram like those in bar charts.

```
>> hist(x)
>> hold on
>> xlabel('Heights')
>> ylabel('Number of People')
```

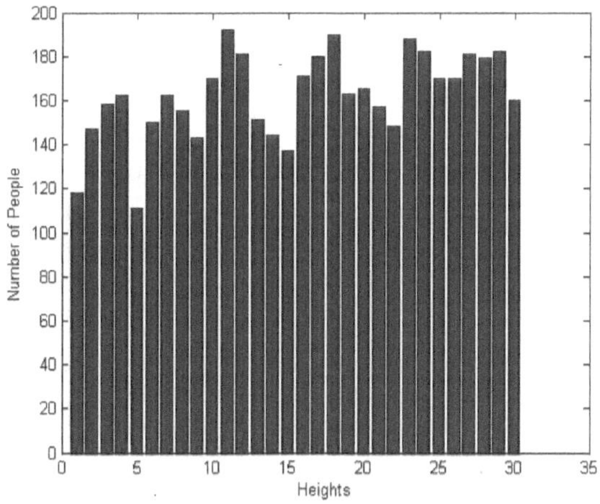

Figure 13.1: Bar Chart for the Example

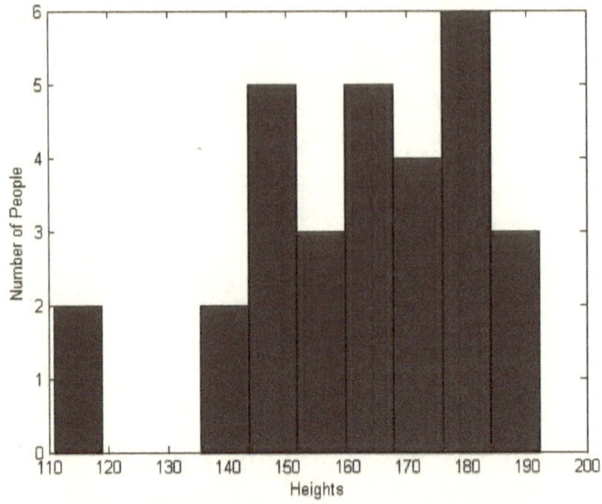

Figure 13.2: Histogram for the Example

The cumulative frequence plot of the data is obtained using the MATLAB command `cdfplot` as follows. The resulting graph is shown in Figure 13.3. The cumulative frequency plot is another type of histogram. The height of the first bar represents the proportion of data points that fall into the given value. The height of the second bar is the frequency or proportion of data points that fall into that given data plus those that fall into the first bar. Thus, any given bar represents the frequency or proportion of data values that fall into that value or anything less than that value.

```
>> cdfplot(x)
```

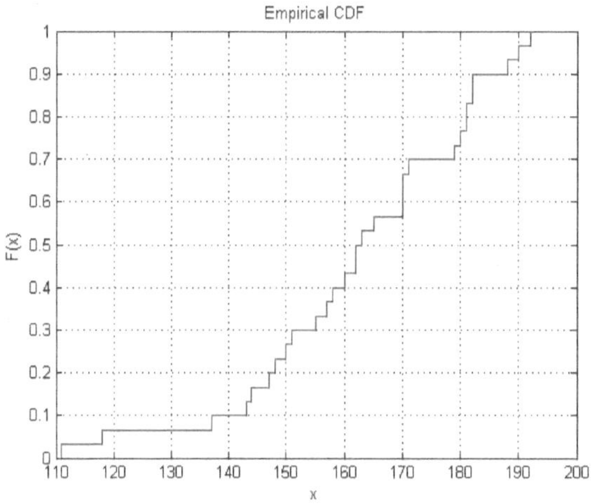

Figure 13.3: Cumulative Frequencey Plot for the Example

The scatterplot of the data is obtained using the MATLAB command `scatter` as follows. The resulting graph is shown in Figure 13.4. But first we need to store the number of data points in a variable (called `y` here) in order to plot them on the x-axis while the actual data points are plotted on the y-axis. Note that the scatterplot shows the heights (or data values) on the y-axis as points on the x-y graph.

```
>> size(x)

ans =

     1    30

>> y = 1:30:1

y =

     1
```

```
>> y = 1:30

y =

  Columns 1 through 11

     1          2          3          4          5          6
7      8          9     10     11

  Columns 12 through 22

    12         13         14         15         16         17
18     19     20     21     22

  Columns 23 through 30

    23         24         25         26         27         28
29     30

>> scatter(y,x)
>> hold on
>> ylabel('Heights')
>> xlabel('Number')
>> scatter(y,x)
```

The pie chart of the data is obtained using the MATLAB command pie as follows. The resulting graph is shown in Figure 13.5. Pie charts show the whole group or population as a circle or pie. The slices of the pie represent each category or subgroup. The sizes of the slices are proportional to the fraction of the whole that each category or subgroup represents. In this example, no attempt is made to group the data in categories or subgroups. The data is plotted into a pie chart as they are given.

```
>> pie(x)
```

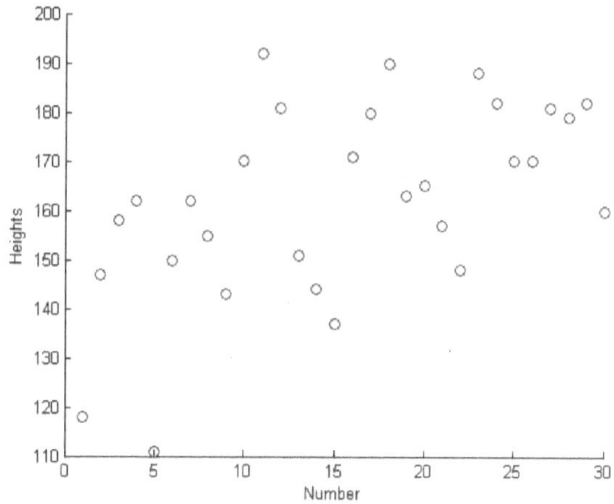

Figure 13.4: Scatterplot of the Data Points of the Example

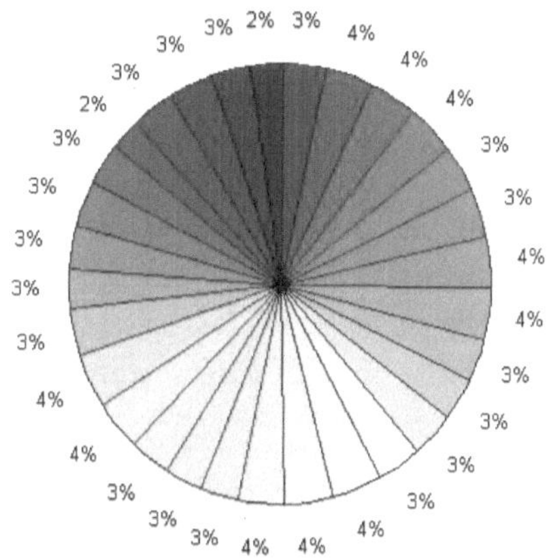

Figure 13.5: Pie Chart of the Data Points of the Example

The rose diagram (some sort of an angular histogram) of the data is obtained using the MATLAB command rose as follows. The resulting graph is shown in Figure 13.6. The rose diagram is circular like the pie chart but shows angular bars instead of slices.

```
>> rose(x)
```

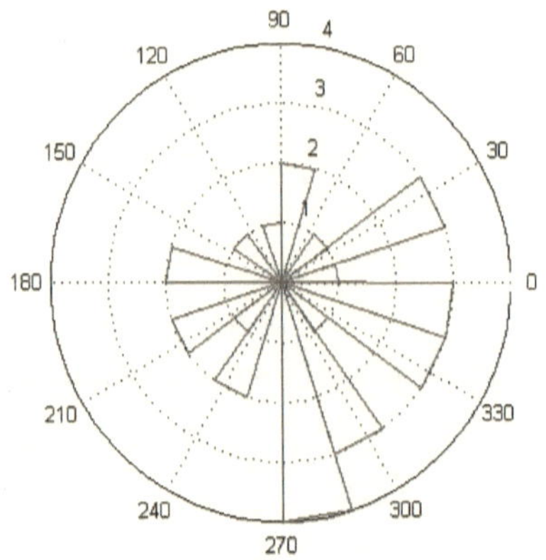

Figure 13.6: Rose Diagram for the Example

A stem plot of the data is obtained using the MATLAB command stem as follows. The resulting graph is shown in Figure 13.7. A stem plot is like a histogram or bar chart, but uses stems (vertical lines) instead of bars.

```
>> stem(x)
>> hold on
>> xlabel('Heights')
>> ylabel('Number of People')
```

A scatterplot of the zscores can also be plotted. Calculation of the zscores was shown previously in Chapter 12. This is repeated here and followed by using the MATLAB command scatter to plot the scatterplot of the zscores. The detailed MATLAB commands follow. The resulting graph is shown in Figure 13.8.

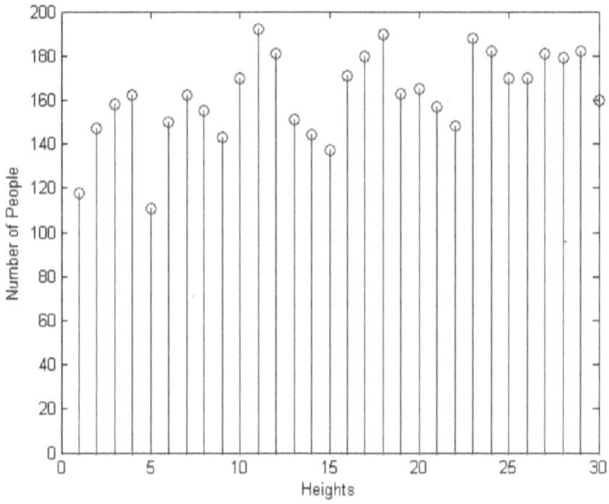

Figure 13.7: A Stem Plot of the Data Points of the Example

```
>> zscore(x)

ans =

  Columns 1 through 7

   -2.2307    -0.7682    -0.2135    -0.0118
-2.5837   -0.6169   -0.0118

  Columns 8 through 14

   -0.3648    -0.9699     0.3917     1.5011
0.9464   -0.5665   -0.9195

  Columns 15 through 21

   -1.2725     0.4421     0.8960     1.4003
0.0387    0.1395   -0.2639
```

```
  Columns 22 through 28

   -0.7178      1.2994      0.9968      0.3917
0.3917    0.9464    0.8455

  Columns 29 through 30

    0.9968   -0.1126

>> scatter(y,ans)
>> hold on
>> xlabel('Number')
>> ylabel('zscore')
```

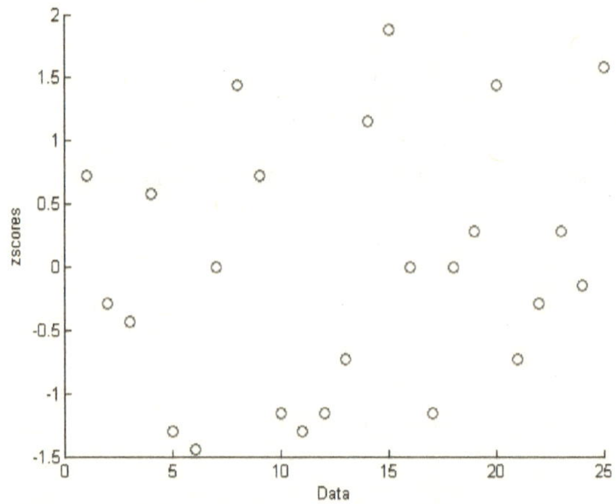

Figure 13.8: Scatterplot of Zscores for the Example

A boxplot (also called a box or whiskers plot) of the data points can also be plotted. This shown in MATLAB below using the command boxplot. The resulting graph is shown in Figure 13.9. A box plot graphically shows the distribution of data by focusing on five key measures: the minimum value, the first quartile(25th percentile), the median (50th percentile), the third quartile (75th percentile), and the maximum value.

```
>> boxplot(x)
>> hold on
>> ylabel('Heights')
```

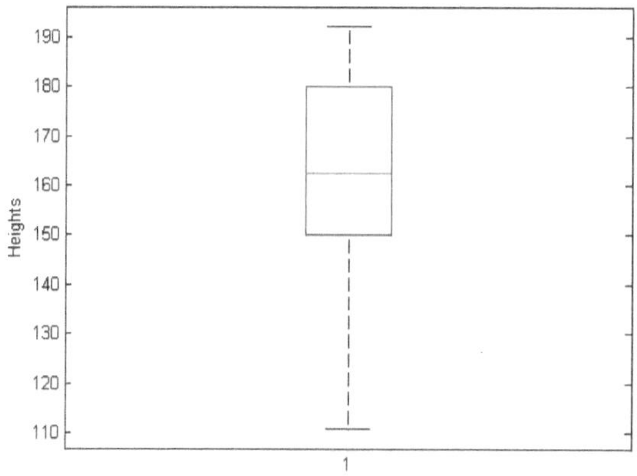

Figure 13.9: A Boxplot of the Data Points of the Example

Exercises

Solve Exercies 1 - 9 using the following set of 25 data points using MATLAB:

25	18	17	24	11
10	20	30	25	12
11	12	15	28	33
20	12	20	22	30
15	18	22	19	31

1. Draw a bar chart based on the above set of data points using MALTAB.

2. Draw a histogram based on the above set of data points using MALTAB.

3. Draw a cumulative frequency plot based on the above set of data points using MATLAB.

4. Draw a scatterplot using the data points in the Exercise above.

5. Draw a pie chart based on the data set in the Exercise above.

6. Draw a rose diagram based on the data set in the Exercise above.

7. Draw a stem plot based on the data points of the Exercise above.

8. Draw a scatterplot of the zscores of the data points of the Exercise above.

9. Draw a boxplot for the Exercise above.

8. Predictive Modeling Fundamentals

This chapter introduces regression analysis as a foundational predictive modeling technique in data science. Regression is presented not only as a statistical tool, but as a core method for learning relationships from data, building predictive models, and performing data-driven inference. This chapter establishes the conceptual link between data, models, and prediction systems.

Regression analysis is a statistical method which helps to determine an approximate form of the relationship between variables. In this chapter we consider the relationship between two variables using simple linear regression[17]. The relationship between three or more variables is presented in Chapter 15 using multiple regression.

Two variables are related linearly if their relationship can be expressed by the equation of a straight line. This is the essence of simple linear regression. The resulting straight line is considered to be a best-fit for the data under study. We assume the equation of the straight line to be written in the following form:

---

[17] There are some specialized MATLAB commands like the command polyfit that can be used to find the best fitting polynomial or straight line for a give set of data. But such specialized and advanced commands are not used in this book. Instead use is made of the various MATLAB commands found in the basic package to solve these types of problems. In addition, using simple basic commands from scratch enhances the learning process considerably.

$$y = a + bx$$

where a and b are constant coefficients to be determined for a specific set of data (the coefficient b represents the slope or gradient of the resulting straight line while the coefficient a represents the y-intercept, i.e. the point where the resulting straight line intersects the y-axis). We will proceed with the detailed derivation of the equations of simple linear regression.

For a set of data x_i and y_i (where $i = 1,2,3,......n$), the above equation becomes for each point (x_i, y_i):

$$y_i = a + bx_i$$
(A)

It should be noted that the above equations is repeated n times, where n is the number of data points. Let us write the above equations for the first few data points as $i = 1,2,3,......n$:

$$y_1 = a + bx_1$$

$$y_2 = a + bx_2$$

$$y_3 = a + bx_3$$

....

....

....

$$y_n = a + bx_n$$

Let us now add the above list of equations to obtain the following expression:

$$y_1 + y_2 + y_3 + ... + y_n = a + bx_1 + a + bx_2 + a + bx_3 + ... + a + bx_n$$

The above equation can be written in the following simplified form:

$$y_1 + y_2 + y_3 + ... + y_n = (a + a + a + ... + a) + (bx_1 + bx_2 + bx_3 + ... + bx_n)$$

where the sum of the a's in the first parenthesis is n times. The above equation can be written using the summation sign $\sum$ as follows:

$$\sum_{i=1}^{n} y_i = na + b\sum_{i=1}^{n} x_i$$

(B)

where the following summation notation is used in equation (B):

$$\sum_{i=1}^{n} y_i = y_1 + y_2 + y_3 + ... + y_n$$

$$na = a + a + a + ... + a \qquad (summed\ \ n\ \ times)$$

$$b\sum_{i=1}^{n} x_i = b(x_1 + x_2 + x_3 + \ldots\ldots + x_n) = bx_1 + bx_2 + bx_3 + \ldots. + bx_n$$

The next step would be to multiply equation (A) by x_i to obtain the following equation:

$$x_i y_i = ax_i + bx_i x_i$$

(C)

Writing the above equation out for each value $i = 1, 2, 3, \ldots\ldots n$, we obtain:

$$x_1 y_1 = ax_1 + bx_1^2$$

$$x_2 y_2 = ax_2 + bx_2^2$$

$$x_3 y_3 = ax_3 + bx_3^2$$

………………

………………

………………

$$x_n y_n = ax_n + bx_n^2$$

Adding the above list of equations results in the following expression:

$$x_1y_1 + x_2y_2 + x_3y_3 + \ldots\ldots + x_ny_n = ax_1 + bx_1^2 + ax_2 + bx_2^2 + ax_3 + bx_3^2 + \ldots\ldots\ldots + ax_n + bx_n^2$$

The above equation can be simplified as follows:

$$x_1y_1 + x_2y_2 + x_3y_3 + \ldots\ldots + x_ny_n = ax_1 + ax_2 + ax_3 + \ldots\ldots + ax_n + bx_1^2 + bx_2^2 + bx_3^2 + \ldots\ldots + bx_n^2$$

The above equation can be written using the summation sign $\sum$ as follows:

$$\sum_{i=1}^{n} x_iy_i = a\sum_{i=1}^{n} x_i + b\sum_{i=1}^{n} x_i^2$$

(D)

where the following summation notation is used in equation (D):

$$\sum_{i=1}^{n} x_iy_i = x_1y_1 + x_2y_2 + x_3y_3 + \ldots\ldots + x_ny_n$$

$$a\sum_{i=1}^{n} x_i = a\left(x_1 + x_2 + x_3 + \ldots\ldots + x_n\right) = ax_1 + ax_2 + ax_3 + \ldots\ldots + ax_n$$

$$b\sum_{i=1}^{n} x_i^2 = b\left(x_1^2 + x_2^2 + x_3^2 + \ldots\ldots + x_n^2\right) = bx_1^2 + bx_2^2 + bx_3^2 + \ldots\ldots + bx_n^2$$

Next we solve equations (B) and (D) simultaneously. The two equations are summarized below followed by their solution which was obtained using the method of elimination manually or using MATLAB with the command `solve` as illustrated in Chapter 10 on Solving Equations:

$$\sum_{i=1}^{n} y_i = na + b\sum_{i=1}^{n} x_i$$

$$\sum_{i=1}^{n} x_i y_i = a\sum_{i=1}^{n} x_i + b\sum_{i=1}^{n} x_i^2$$

The solution of the above two linear equations for the constants a and b is obtained as follows:

$$b = \frac{n\sum_{i=1}^{n} x_i y_i - \sum_{i=1}^{n} x_i \sum_{i=1}^{n} y_i}{n\sum_{i=1}^{n} x_i^2 - \left(\sum_{i=1}^{n} x_i\right)^2}$$

(I)

$$a = \frac{\sum_{i=1}^{n} x_i^2 \sum_{i=1}^{n} y_i - \sum_{i=1}^{n} x_i \sum_{i=1}^{n} x_i y_i}{n\sum_{i=1}^{n} x_i^2 - \left(\sum_{i=1}^{n} x_i\right)^2}$$

(II)

The procedure for curve fitting a straight line through the method of simple linear regression is summarized in the following five steps:

1. Calculate the sums $\sum_{i=1}^{n} x_i$, $\sum_{i=1}^{n} y_i$, $\sum_{i=1}^{n} x_i^2$, and $\sum_{i=1}^{n} x_i y_i$ using MATLAB (if calculated manually, it is preferred to do it in a table). Table 14.1 shows each sum above along with the needed MATLAB command to compute it.

Table 14.1: MALTAB Command Needed to Compute Each Sum

Sum	MATLAB Command
x_i	`x`
y_i	`y`
$\sum_{i=1}^{n} x_i$	`sum(x)`
$\sum_{i=1}^{n} y_i$	`sum(y)`
$\sum_{i=1}^{n} x_i^2$	`mxx = x.*x` `sum(mxx)`
$\sum_{i=1}^{n} x_i y_i$	`mxy = x.*y` `sum(mxy)`
$\left(\sum_{i=1}^{n} x_i\right)^2$	`(sum(x))^2`

2. Substitute the above four sums into equations (I) and (II) to obtain the two constants a and b.

3. Using the two values of the constants a and b obtained above, write the equation of the best fitting line as follows:

   $$y = a + bx$$

4. To check the results or to view the results graphically, plot the data using the MATLAB command `scatter`, then plot the best fitting line on the same graph using the MATLAB command `plot`. These two

commands will be illustrated in the numerical example that will follow.

5. This last step is optional. We can calculate various types of errors between the data and the best fitting line, including the maximum error. However, this step will not be illustrated in the numerical example that will follow.

In order to illustrate simple linear regression in this chapter, we employ what is known as the Least Squares procedure. This procedure is used to obtain a best-fit curve or a best-fit line that represents the data based on simple linear regression. This is shown below using an example.

Let us now solve a numerical example using the above five steps. The following data (see Table 14.2) show the monthly salaries (x, in dollars) and the amount of monthly rent payments (y, in dollars) for a random sample of 16 employees at a certain company. Let us obtain the equation of the best fitting line using simple linear regression.

Table 14.2: Data for the Numerical Example

x	600	700	650	680	750	800	900	850
y	150	180	200	230	250	280	300	350

Table 14.2: Data for the Numerical Example (Continued)

x	950	1000	1020	1050	1100	1200	1250	1300
y	400	430	450	480	500	510	550	560

Let us now proceed to start the solution of the numerical example given above using the five steps outlined above for simple linear regression.

Step 1: Calculate the sums $\sum_{i=1}^{n} x_i$, $\sum_{i=1}^{n} y_i$, $\sum_{i=1}^{n} x_i^2$, and $\sum_{i=1}^{n} x_i y_i$ using the MATLAB command `sum`[18] as follows:

1.1: Let us first input the data as follows:

```
>> x = [600 700 650 680 750 800 900
850 950 1000 1020 1050 1100 1200 1250
1300]

x =

  Columns 1 through 6

         600           700           650
680      750           800

  Columns 7 through 12

         900           850           950
1000     1020          1050

  Columns 13 through 16

        1100          1200          1250
1300

>> y = [150 180 200 230 250 280 300
350 400 430 450 480 500 510 550 560]
```

[18] For more details about the MATLAB command `sum`, see Chapter 6 on Vectors.

```
y =

  Columns 1 through 12

   150    180    200    230    250    280
300    350    400    430    450    480

  Columns 13 through 16

   500    510    550    560
```

1.2: Let us calculate the sum $\sum_{i=1}^{n} x_i$:

```
>> sumx = sum(x)

sumx =

       14800
```

1.3: Let us calculate the sum $\sum_{i=1}^{n} y_i$:

```
>> sumy = sum(y)

sumy =

        5820
```

1.4: Let us calculate the sum $\sum_{i=1}^{n} x_i^2$:

```
>> mxx = x.*x

mxx =

  Columns 1 through 5

        360000        490000        422500
462400        562500
```

```
     Columns 6 through 10

         640000         810000         722500
902500      1000000

     Columns 11 through 15

        1040400        1102500        1210000
1440000      1562500

     Column 16

        1690000

   >> sumxx = sum(mxx)

   sumxx =

       14417800
```

1.5: Let us calculate the sum $\sum_{i=1}^{n} x_i y_i$:

```
   >> mxy = x.*y

   mxy =

     Columns 1 through 5

          90000         126000         130000
156400       187500

     Columns 6 through 10

         224000         270000         297500
380000       430000

     Columns 11 through 15

         459000         504000         550000
612000       687500
```

```
  Column 16

      728000

>> sumxy = sum(mxy)

sumxy =

     5831900
```

1.6: Let us calculate the square sum $\left(\sum_{i=1}^{n} x_i\right)^2$:

```
>> sumx2 = sumx*sumx

sumx2 =

   219040000
```

1.7: Let us calculate the number of data points using the `size` command:

```
>> n = size(x)

n =

     1    16

>> n1 = n(2)

n1 =

    16
```

Step 2: Substitute the values of the above four sums into equations (I) and (II), evaluate the two expressions in these two equations, and obtain the values of the constants a and b. This is performed in MATLAB as follows:

```
>> a = (sumxx*sumy - sumx*sumxy)/(n1*sumxx - sumx2)

a =

 -206.1456

>> b = (n1*sumxy - sumx*sumy)/(n1*sumxx - sumx2)

b =

    0.6161
```

Step 3: Using the values of the constants a and b obtained above, we can finally write the equation of the best fitting line as follows:

$$y = -206.1456 + 0.6161x$$

Step 4: In order to compare the best fitting line with the data given, let us know plot the data along with the best fitting line on the same graph. This is performed using the two MATLAB commands `scatter` and `plot` as follows:

4.1: Let us calculate the y values using the best fit equation above as follows:

```
>> y1 = a + b*x

y1 =

  Columns 1 through 7

  163.5164    225.1268    194.3216
212.8047  255.9319  286.7371  348.3474

  Columns 8 through 14
```

```
    317.5423       379.1526       409.9577
422.2798  440.7629  471.5681  533.1784

    Columns 15 through 16

    563.9836  594.7887
```

4.2: Let us plot the best fitting straight line:

```
>> plot(x,y1)
>> hold on
```

4.3: Let us plot the given data points:

```
>> scatter(x,y)
```

4.4: Let us label the x-axis and the y-axis:

```
>> xlabel('x')
>> ylabel('y')
```

The graph generated by MATLAB is shown in Figure 14.1. It is clear from the figure that the generated best fitting line truly represents the data points given in the example.

Step 5: This error calculation step is optional and is not demonstrated here.

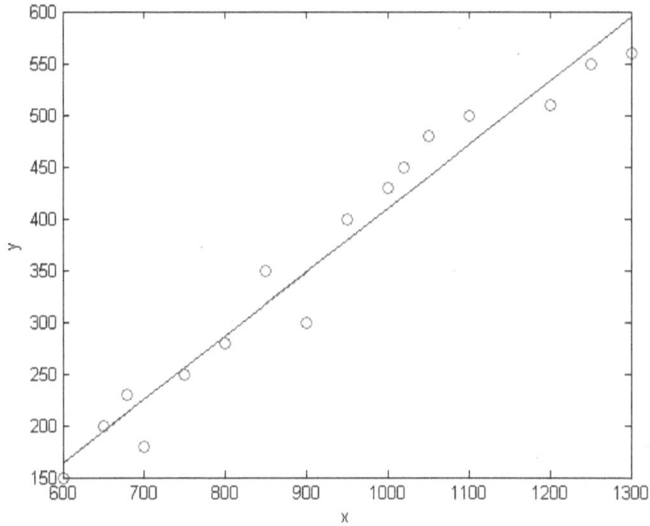

Figure 14.1: Best Fitting Line Along with the Given Data Points for the Example

Exercises

Solve the following three exercises on simple regression analysis. The first two should be done using simple linear regression using the five steps presented in this chapter. However, the third exercise is different. The third exercise is on the topic of simple nonlinear regression (specifically simple quadratic or simple parabolic regression) that is not illustrated in this chapter. However, enough information, hints, and steps are given in the exercise to make its solution attainable by the student.

1. A company makes a study of its assets (x, in thousands of dollars) and its expenditure (y, in thousands of dollars) for a specific department. The following data are given (see Table 14.3):

Table 14.3: Data for Exercise 1

x	50	52	56	61	66	70	80	84	88	92
y	200	220	250	310	360	400	420	440	500	550

Find the equation of the best fitting for the above data. Plot the data along with the best fitting line on the same graph.

2. The following data (see Table 14.4) shows the expenditure x and the net income y, both in dollars, for a random sample of 12 firms.

Table 14.4: Data for Exercise 2

x	10	20	25	31	36	44	55	60	62	65	70	75
y	15	25	30	35	40	50	60	65	70	75	80	85

Find the equation of the best fitting for the above data. Plot the data along with the best fitting line on the same graph.

3. This exercise is optional for the beginning student and involves the equations of simple nonlinear regression, specifically what is called simple quadratic regression (also called simple parabolic regression). We will not solve a numerical problem in this exercise, but only want to write the equations for this special case of simple nonlinear regression.

   For a given set of data (x_i, y_i), assume the equation of the best fitting parabola to be quadratic in the general form:

   $$y = a + bx + cx^2$$

   Derive the three algebraic equations for the constant coefficients a, b, and c using the method of simple nonlinear regression.

9. Data-Driven Modeling and Prediction Systems

This chapter advances regression analysis as a structured modeling framework for data science applications. Emphasis is placed on model construction, interpretation, and validation as components of data-driven systems. Regression is framed as an entry point into machine learning, forecasting, and intelligent analytical systems used across scientific and applied domains.

In this chapter we study multiple regression analysis. In this case, we have linear multiple regression and nonlinear multiple regression. We will derive the equations of linear multiple regression and illustrate how to use them using a numerical example[19].

Linear multiple regression is used to approximate the relationship between three or more variables. Two cases will be illustrated in this chapter: the first case involves three variables while the second case involves four variables. Both cases are based on the least squares technique of statistical analysis.

Consider a set of n data points (x_i, y_i, z_i) in three dimensions. In this case and since we are dealing with three-dimensional space, we obtain a best fitting plane instead of a best fitting line. The

[19] Multiple nonlinear regression is an advanced technique and will not be illustrated in this beginner's book.

equation of the best fitting plane can be written in general as follows:

$$z = a + bx + cy$$

What is required is find the values of the three constants a, b, and c. The same procedure of Chapter 14 is followed here but note that we are dealing with three dimensions instead of two dimensions. The above equation is repeated below with subscript $i = 1, 2, 3, ...n$:

$$z_i = a + bx_i + cy_i$$
(E)

It should be noted that the above equations is repeated n times, where n is the number of data points. Let us write the above equations for the first few data points as $i = 1, 2, 3, ...n$:

$$z_1 = a + bx_1 + cy_1$$

$$z_2 = a + bx_2 + cy_2$$

$$z_3 = a + bx_3 + cy_3$$

…………...

……………

……………

$$z_n = a + bx_n + cy_n$$

Let us now add the above list of equations to obtain the following expression:

$$z_1 + z_2 + z_3 + \dots\dots + z_n = a + bx_1 + cy_1 + a + bx_2 + cy_2 \\ + a + bx_3 + cy_3 + \dots\dots + a + bx_n + cy_n$$

The above equation can be written in the following simplified form:

$$z_1 + z_2 + z_3 + \dots\dots + z_n = (a + a + a + \dots. + a) \\ + (bx_1 + bx_2 + bx_3 + \dots\dots + bx_n) + (cy_1 + cy_2 + cy_3 + \dots\dots + cy_n$$

where the sum of the a's in the first parenthesis is n times. The above equation can be written using the summation sign $\sum$ as follows:

$$\sum_{i=1}^{n} z_i = na + b\sum_{i=1}^{n} x_i + c\sum_{i=1}^{n} y_i$$

(F)

where the following summation notation is used in equation (F):

$$\sum_{i=1}^{n} z_i = z_1 + z_2 + z_3 + \dots\dots + z_n$$

$$na = a + a + a + \dots\dots + a \qquad (\textit{summed } n \textit{ times})$$

$$b\sum_{i=1}^{n} x_i = b(x_1 + x_2 + x_3 + \ldots\ldots + x_n) = bx_1 + bx_2 + bx_3 + \ldots. + bx_n$$

$$c\sum_{i=1}^{n} y_i = c(y_1 + y_2 + y_3 + \ldots\ldots + y_n) = cy_1 + cy_2 + cy_3 + \ldots. + cy_n$$

The next step would be to multiply equation (E) by x_i to obtain the following equation:

$$x_i z_i = ax_i + bx_i x_i + cx_i y_i$$
(G)

Writing the above equation out for each value $i = 1, 2, 3, \ldots\ldots n$, we obtain:

$$x_1 z_1 = ax_1 + bx_1^2 + cx_1 y_1$$

$$x_2 z_2 = ax_2 + bx_2^2 + cx_2 y_2$$

$$x_3 z_3 = ax_3 + bx_3^2 + cx_3 y_3$$

....

....

....

$$x_n z_n = ax_n + bx_n^2 + cx_n y_n$$

Adding the above list of equations results in the following expression:

$$x_1z_1 + x_2z_2 + x_3z_3 + ... + x_nz_n = ax_1 + bx_1^2 + cx_1y_1$$
$$+ ax_2 + bx_2^2 + cx_2y_2 + ax_3 + bx_3^2 + cx_3y_3 + ... + ax_n + bx_n^2 + cx$$

The above equation can be simplified as follows:

$$x_1z_1 + x_2z_2 + x_3z_3 + ... + x_nz_n = ax_1 + ax_2 + ax_3 + ... + ax_n$$
$$+ bx_1^2 + bx_2^2 + bx_3^2 + ... + bx_n^2 + cx_1y_1 + cx_2y_2 + cx_3y_3 + ... + c$$

The above equation can be written using the summation sign $\sum$ as follows:

$$\sum_{i=1}^{n} x_iz_i = a\sum_{i=1}^{n} x_i + b\sum_{i=1}^{n} x_i^2 + c\sum_{i=1}^{n} x_iy_i$$

(H)

where the following summation notation is used in equation (H):

$$\sum_{i=1}^{n} x_iz_i = x_1z_1 + x_2z_2 + x_3z_3 + ... + x_nz_n$$

$$a\sum_{i=1}^{n} x_i = a\left(x_1 + x_2 + x_3 + ... + x_n\right) = ax_1 + ax_2 + ax_3 + ... + a$$

$$b\sum_{i=1}^{n} x_i^2 = b\left(x_1^2 + x_2^2 + x_3^2 + \ldots\ldots + x_n^2\right) = bx_1^2 + bx_2^2 + bx_3^2 + \ldots\ldots + bx_n^2$$

$$c\sum_{i=1}^{n} x_i y_i = c\left(x_1 y_1 + x_2 y_2 + x_3 y_3 + \ldots\ldots + x_n y_n\right)$$

$$= cx_1 y_1 + cx_2 y_2 + cx_3 y_3 + \ldots\ldots + cx_n y_n$$

The next step would be to multiply equation (E) by y_i to obtain the following equation:

$$y_i z_i = ay_i + bx_i y_i + cy_i y_i$$
(J)

Writing the above equation out for each value $i = 1, 2, 3, \ldots\ldots n$, we obtain:

$$y_1 z_1 = ay_1 + bx_1 y_1 + cy_1^2$$

$$y_2 z_2 = ay_2 + bx_2 y_2 + cy_2^2$$

$$y_3 z_3 = ay_3 + bx_3 y_3 + cy_3^2$$

....

....

....

$$y_n z_n = ay_n + bx_n y_n + cy_n^2$$

Adding the above list of equations results in the following expression:

$$y_1z_1 + y_2z_2 + y_3z_3 + ... + y_nz_n = ay_1 + bx_1y_1 + cy_1^2 + ay_2 + bx_2y_2 + cy_2^2 + ax_3 + bx_3y_3 + cy_3^2 + ... + ay_n + bx_ny_n + c$$

The above equation can be simplified as follows:

$$y_1z_1 + y_2z_2 + y_3z_3 + ... + y_nz_n = ay_1 + ay_2 + ay_3 + ... + ay_n + bx_1y_1 + bx_2y_2 + bx_3y_3 + ... + bx_ny_n + cy_1^2 + cy_2^2 + cy_3^2 + ... + c$$

The above equation can be written using the summation sign $\sum$ as follows:

$$\sum_{i=1}^{n} y_iz_i = a\sum_{i=1}^{n} y_i + b\sum_{i=1}^{n} x_iy_i + c\sum_{i=1}^{n} y_i^2$$

(K)

where the following summation notation is used in equation (H):

$$\sum_{i=1}^{n} y_iz_i = y_1z_1 + y_2z_2 + y_3z_3 + ... + y_nz_n$$

$$a\sum_{i=1}^{n} y_i = a(y_1 + y_2 + y_3 + ... + y_n) = ay_1 + ay_2 + ay_3 + ... + a$$

$$b\sum_{i=1}^{n} x_i y_i = b\left(x_1 y_1 + x_2 y_2 + x_3 y_3 + \ldots\ldots + x_n y_n\right)$$

$$= bx_1 y_1 + bx_2 y_2 + bx_3 y_3 + \ldots\ldots + bx_n y_n$$

$$c\sum_{i=1}^{n} y_i^2 = c\left(y_1^2 + y_2^2 + y_3^2 + \ldots\ldots + y_n^2\right) = cy_1^2 + cy_2^2 + cy_3^2 + \ldots\ldots + cy_n^2$$

Next we list equations (F), (H) and (K) as shown in the boxed list below. These three equations should be solved simultaneously to obtain the three constants a, b, and c.

It is not advised to solve the three equations below symbolically and obtain a closed-form or analytical set of expressions for the three constants a, b, and c. Instead, it is advised to write out the three equations above explicitly for each specific set of data, then solve the three equations numerically using the solutions techniques illustrated in Chapter 10 on Solving Equations.

$$\sum_{i=1}^{n} z_i = na + b\sum_{i=1}^{n} x_i + c\sum_{i=1}^{n} y_i$$

(III)

$$\sum_{i=1}^{n} x_i z_i = a\sum_{i=1}^{n} x_i + b\sum_{i=1}^{n} x_i^2 + c\sum_{i=1}^{n} x_i y_i$$

(IV)

$$\sum_{i=1}^{n} y_i z_i = a\sum_{i=1}^{n} y_i + b\sum_{i=1}^{n} x_i y_i + c\sum_{i=1}^{n} y_i^2$$

(V)

The three equations above can be written out explicitly for each specific set of data points by calculating the various sums that appear in the equations. There are eight sums to be calculated in each case; namely $\sum_{i=1}^{n} x_i$, $\sum_{i=1}^{n} y_i$, $\sum_{i=1}^{n} z_i$, $\sum_{i=1}^{n} x_i y_i$, $\sum_{i=1}^{n} x_i z_i$, $\sum_{i=1}^{n} y_i z_i$, $\sum_{i=1}^{n} x_i^2$, and $\sum_{i=1}^{n} y_i^2$.

The procedure for curve fitting a plane through the method of multiple linear regression is summarized in the following four steps (one step from Chapter 14 is obviously missing; the step of showing the data and the best fit on a graph. It is not easy to show these data along with the best fit on a graph because we are now dealing with data in three-dimensional space):

1. Calculate the sums $\sum_{i=1}^{n} x_i$, $\sum_{i=1}^{n} y_i$, $\sum_{i=1}^{n} z_i$, $\sum_{i=1}^{n} x_i y_i$, $\sum_{i=1}^{n} x_i z_i$, $\sum_{i=1}^{n} y_i z_i$, $\sum_{i=1}^{n} x_i^2$, and $\sum_{i=1}^{n} y_i^2$ using MATLAB (if calculated manually, it is preferred to do it in a table). Table 15.1 shows each sum along with needed MATLAB command to compute it.

Table 15.1: MALATB Command Needed to Compute Each Sum

Sum	MATLAB Command
x_i	`X`
y_i	`y`
z_i	`z`
$\sum_{i=1}^{n} x_i$	`sum(x)`
$\sum_{i=1}^{n} y_i$	`sum(y)`
$\sum_{i=1}^{n} z_i$	`sum(z)`
$\sum_{i=1}^{n} x_i^2$	`mxx = x.*x` `sum(mxx)`
$\sum_{i=1}^{n} y_i^2$	`myy = y.*y` `sum(myy)`
$\sum_{i=1}^{n} x_i y_i$	`mxy = x.*y` `sum(mxy)`
$\sum_{i=1}^{n} x_i z_i$	`mxz = x.*z` `sum(mxz)`
$\sum_{i=1}^{n} y_i z_i$	`myz = y.*z` `sum(myz)`

2. Substitute the above eight sums into equations (III), (IV), and (V), and solve them, in order to obtain the three constants a, b, and c.

3. Using the three values of the constants a, b, and c obtained above, write the equation of the best fitting plane as follows:

$$z = a + bx + cy$$

4. This last step is optional. We can calculate various types of errors between the data and the best fitting plane, including the maximum error. However, this step will not be illustrated in the numerical example that will follow.

Let us now solve a numerical example using the above four steps. The following data (see Table 15.2) is given in three dimensions. Let us obtain the equation of the best fitting plane using multiple linear regression.

Table 15.2: Data in Three Dimensions for the Numerical Example

x	120	150	210	290	350	400	500
y	10	12	17	25	30	33	38
z	1050	2000	2100	1500	2100	3300	1100

The solution of this numerical example using MATLAB follows the four steps outlined above and is shown in detail below:

Step 1: Let us calculate the eight sums $\sum_{i=1}^{n} x_i$, $\sum_{i=1}^{n} y_i$, $\sum_{i=1}^{n} z_i$, $\sum_{i=1}^{n} x_i y_i$, $\sum_{i=1}^{n} x_i z_i$, $\sum_{i=1}^{n} y_i z_i$, $\sum_{i=1}^{n} x_i^2$, and $\sum_{i=1}^{n} y_i^2$ for the data given in Table 4 using MATLAB. The two basic MATLAB commands to be used are the command `sum` (summation command) and the command `.*` (term-by-term vector multiplication command). Both these commands were studied in great detail in Chapter 6 on vectors.

1.1: Let us enter the data points given in the Example:

```
>> x = [120 150 210 290 350 400 500]

x =

   120    150    210    290    350    400
500

>> y = [10 12 17 25 30 33 38]

y =

    10     12     17     25     30     33
38

>> z = [1050 2000 2100 1500 2100 3300
1100]

z =

  Columns 1 through 6
```

```
        1050              2000              2100
    1500          2100              3300

  Column 7

        1100
```

1.2: Let us calculate the sum $\sum_{i=1}^{n} x_i$:

```
>> sumx = sum(x)

sumx =

        2020
```

1.3: Let us calculate the sum $\sum_{i=1}^{n} y_i$:

```
>> sumy = sum(y)

sumy =

   165
```

1.4: Let us calculate the sum $\sum_{i=1}^{n} z_i$:

```
>> sumz = sum(z)

sumz =

       13150
```

1.5: Let us calculate the sum $\sum_{i=1}^{n} x_i^2$:

```
>> mxx = x.*x
```

```
mxx =

  Columns 1 through 5

        14400          22500          44100
84100        122500

  Columns 6 through 7

       160000       250000

>> sumxx = sum(mxx)

sumxx =

       697600
```

1.6: Let us calculate the sum $\sum_{i=1}^{n} x_i y_i$:

```
>> mxy = x.*y

mxy =

  Columns 1 through 5

         1200           1800           3570
7250          10500

  Columns 6 through 7

        13200        19000

>> sumxy = sum(mxy)

sumxy =

        56520
```

1.7: Let us calculate the sum $\sum_{i=1}^{n} x_i z_i$:

```
>> mxz = x.*z

mxz =

  Columns 1 through 5

      126000          300000          441000
435000        735000

  Columns 6 through 7

     1320000        550000

>> sumxz = sum(mxz)

sumxz =

     3907000
```

1.8: Let us calculate the sum $\sum_{i=1}^{n} y_i z_i$:

```
>> myz = y.*z

myz =

  Columns 1 through 5

        10500           24000           35700
37500         63000

  Columns 6 through 7

      108900         41800

>> sumyz = sum(myz)
```

```
sumyz =

      321400
```

1.9: Let us calculate the sum $\sum_{i=1}^{n} y_i^2$:

```
>> myy = y.*y

myy =

  Columns 1 through 5

         100              144              289
625           900

  Columns 6 through 7

        1089          1444

>> sumyy = sum(myy)

sumyy =

        4591
```

1.10: Let us calculate the number of data points given in the Example using the `size` command:

```
>> n = size(x)

n =

     1     7

>> n1 = n(2)

n1 =
```

```
      7
```

Step 2: Write out the three equations (III), (IV), and (V) explicitly for the set of data given in Table 4 using the results of Step 1 above. Then solve the three resulting equations simultaneously for the constants a, b, and c using the MATLAB command `solve` that was illustrated in detail in Chapter 10 on Solving Equations.

The system of three simultaneous algebraic equations (III), (IV), and (V) can now be written for this example as follows:

$$7a + 2020b + 165c = 13150$$

$$2020a + 697600b + 56520c = 3907000$$

$$165a + 56520b + 4591c = 321400$$

Next, the above system of three simultanaeous algebraic equation is solve using the inverse matrix method in MATLAB as follows (check Chapter 10 on Solving Equations for details):

```
>> A = [n1 sumx sumy ; sumx sumxx
sumxy ; sumy sumxy sumyy]

A =

           7        2020         165
        2020      697600       56520
         165       56520        4591

>> b = [sumz ; sumxz ; sumyz]

b =
```

```
        13150
      3907000
       321400

>> inv(A)*b

ans =

   1.0e+03 *

    1.2850
   -0.0198
    0.2674

>> a = ans(1)

a =

   1.2850e+03

>> b = ans(2)

b =

  -19.7884

>> c = ans(3)

c =

  267.4393
```

Step 3: Using the values of the constants a, b, and c obtained from Step 2 above, write the equation of the best fitting plane as follows:

$$z = 1285 - 19.7884x + 267.4393y$$

Let us finally try to make a 3D plot of the best fitting plane along with the give data points. This is performed using MATLAB as follows:

```
>> z1 = a + b*x + c*y

z1 =

   1.0e+03 *

    1.5848          1.5260          1.6759
2.2324     2.3822     2.1951     1.5535

>> scatter(x,y,z)
>> hold on
>> plot3(x,y,z1)
>> xlabel('x')
>> ylabel('y')
>> zlabel('z')
```

Note that the MATLAB command `plot3` instead of the command `plot` is used above. The resulting graph is shown in Figure 15.1.

Step 4: This step of calculating errors is optional and is not illustrated for this example.

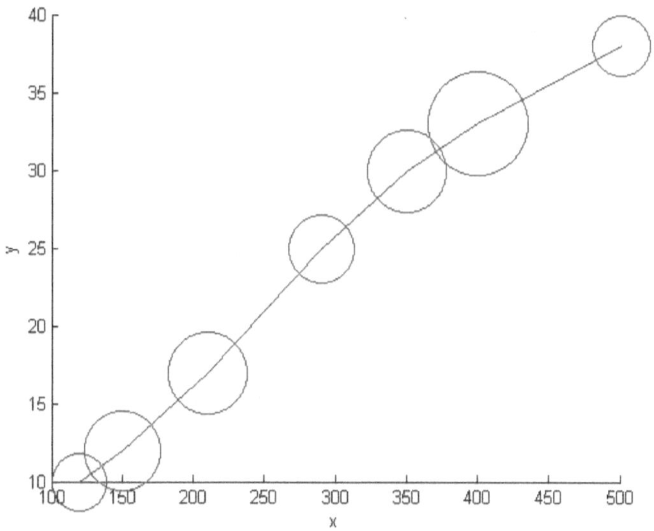

Figure 15.1: Best Fitting Plane and the Given Date Points for the Example

Next, we discuss what needs to be done if one wants to perform multiple linear regression on a set of data consisting of four variables in four-dimensional space. Given a set of n data points (x_i, y_i, z_i, w_i), one can follow the procedure of derivation given above to obtain the equation of the best fitting hyperplane in four-dimensional space. The equation of the best fitting hyperplane can be written in general as follows:

$$w = a + bx + cy + dz$$

We will not provide the details of the derivation, but list only the four simultaneous equations that need to be solved to obtain the four constants a, b, c, and d. The necessary four simultaneous equations are listed in the box below:

$$\sum_{i=1}^{n} w_i = na + b\sum_{i=1}^{n} x_i + c\sum_{i=1}^{n} y_i + d\sum_{i=1}^{n} z_i$$
(VI)

$$\sum_{i=1}^{n} x_i w_i = a\sum_{i=1}^{n} x_i + b\sum_{i=1}^{n} x_i^2 + c\sum_{i=1}^{n} x_i y_i + d\sum_{i=1}^{n} x_i z_i$$
(VII)

$$\sum_{i=1}^{n} y_i w_i = a\sum_{i=1}^{n} y_i + b\sum_{i=1}^{n} x_i y_i + c\sum_{i=1}^{n} y_i^2 + d\sum_{i=1}^{n} y_i z_i$$
(VIII)

$$\sum_{i=1}^{n} z_i w_i = a\sum_{i=1}^{n} z_i + b\sum_{i=1}^{n} x_i z_i + c\sum_{i=1}^{n} y_i z_i + d\sum_{i=1}^{n} z_i^2$$
(IX)

Next, if one considers the relationships between five variables in five-dimensional space, then one needs to write the equation of the best fitting hyperplane in five-dimensional space, then write a set of five simultaneous equations from which one can obtain the set of five constants needed for the equation of the hyperplane. This can be done by following the same procedure illustrated in this chapter for three variables.

Finally, the equations of multiple linear regression have been studied in detail in this chapter. On the other hand, the equations of multiple nonlinear regression are too complicated and are beyond the scope of this beginner's book. The interested reader is referred to the updated list of references at the end of this book for more details.

Exercises

Solve Exercises 1 and 2 following the four steps outlined in this chapter for multiple linear regression. Exercise 3 should be attempted only by advanced readers who are familiar with the derivation of the equations of multiple linear regression.

1. The following data (see Table 15.3) is give for three variables in three-dimensional space. Obtain the equation of the best fitting plane in three dimensions using multiple linear regression. Use MATLAB as illustrated in this chapter. (*Hint*: Use equations III, IV, and V).

Table 15.3: Data in Three Dimensions for Exercise 1

x	109	118	124	100	90	120	150
y	450	440	500	600	450	550	600
z	11	16	17	20	22	26	31

2. The following data (see Table 15.4) is give for four variables in four-dimensional space. Obtain the equation of the best fitting hyperplane in four dimensions using multiple linear regression. Use MATLAB as illustrated in this chapter.

(*Hint*: Use equations VI, VII, VIII, and IX).

Table 15.4: Data in Three Dimensions for Exercise 2

x	234	245	257	270	250	270	299
y	2	4	7	9	7	11	13
z	890	810	910	900	110 0	120 0	100 0
w	103 0	110 0	120 0	135 0	144 0	150 0	157 0

3. Derive equations VI, VII, VIII, and XI in detail using multiple linear regression for four variables in four dimensions. This exercise should be attempted only by advanced readers (*Hint*: The detailed derivation can be found in the Solutions Manual at the end of this book).

Solutions to Exercises

1. Chapter 1

1. Perform the operation 3*4+6. The order of the operations will be discussed in subsequent chapters.

```
>> 3*4+6

ans =

    18
```

2. Perform the operation $\cos(5)$. The value of 5 is in radians.

```
>> cos(5)

ans =

    0.2837
```

3. Perform the operation $3\sqrt{6+x}$ for $x = 4$.

```
>> x = 4

x =

     4

>> 3*sqrt(6+x)

ans =
```

```
     9.4868
```

4. Assign the value of 5.2 to the variable y.

```
>> y = 5.2

y =

     5.2000
```

5. Assign the values of 3 and 4 to the variables x and y, respectively, then calculate the value of z where $z = 2x - 7y$.

```
>> x = 3;
>> y = 4;
>> z = 2*x - 7*y

z =

   -22
```

6. Obtain help on the inv command.

```
>> help inv
 INV    Matrix inverse.
    INV(X) is the inverse of the
square matrix X.
    A warning message is printed
if X is badly scaled or
    nearly singular.

    See also slash, pinv, cond,
condest, lsqnonneg, lscov.
```

```
    Overloaded functions or
methods (ones with the same name
in other directories)
       help sym/inv.m

    Reference page in Help
browser
       doc inv
```

7. Generate the following matrix: A

$$A = \begin{bmatrix} 1 & 0 & 2 & -3 \\ 0 & 5 & 2 & 2 \\ 1 & 2 & 3 & 4 \\ -2 & 0 & 1 & 3 \end{bmatrix}$$

```
>> A = [1 0 2 -3 ; 0 5 2 2 ; 1 2 3 4 ; -2 0 1 3]

A =

     1     0     2    -3
     0     5     2     2
     1     2     3     4
    -2     0     1     3
```

8. Generate the following vector b

$$b = \begin{Bmatrix} 1 \\ 2 \\ 3 \\ 4 \end{Bmatrix}$$

```
>> b = [1 ; 2 ; 3 ; 4]
```

```
b =

     1
     2
     3
     4
```

9. Evaluate the vector c where $\{c\} = [A]\{b\}$ where A is the matrix given in Exercise 7 above and b is the vector given in Exercise 8 above.

```
>> c = A*b

c =

    -5
    24
    30
    13
```

10. Solve the following system of simultaneous algebraic equation using Gaussian elimination.

$$\begin{bmatrix} 5 & 2 \\ 1 & 3 \end{bmatrix} \begin{Bmatrix} x_1 \\ x_2 \end{Bmatrix} = \begin{Bmatrix} 3 \\ -1 \end{Bmatrix}$$

```
>> A = [ 5 2 ; 1 3]

A =

     5     2
     1     3

>> b = [3 ; -1]

b =
```

```
     3
    -1

>> x = A\b

x =

    0.8462
   -0.6154
```

11. Solve the system of simultaneous algebraic equations of Exercise 10 above using matrix inversion.

```
>> x = inv(A)*b

x =

    0.8462
   -0.6154
```

12. Generate the following matrix X:

$$X = \begin{bmatrix} 1 & 0 & 6 \\ 1 & 2 & 3 \\ 4 & 5 & -2 \end{bmatrix}$$

```
>> X = [1 0 6 ; 1 2 3 ; 4 5 -2]

X =

     1     0     6
     1     2     3
     4     5    -2
```

13. Extract the sub-matrix in rows 2 to 3 and columns 1 to 2 of the matrix X in Exercise 12 above.

```
>> X(2:3,1:2)

ans =

     1     2
     4     5
```

14. Extract the second column of the matrix X in Exercise 12 above.

```
>> X(1:3,2)

ans =

     0
     2
     5
```

15. Extract the first row of the matrix X in Exercise 12 above.

```
>> X(1,1:3)

ans =

     1     0     6
```

16. Extract the element in row 1 and column 3 of the matrix X in Exercise 12 above.

```
>> X(1,3)

ans =

     6
```

17. Generate the row vector x with integer values ranging from 1 to 9.

```
>> x = [1 2 3 4 5 6 7 8 9]

x =

        1         2         3         4         5
6       7         8         9
```

18. Plot the graph of the function $y = x^3 - 2$ for the range of the values of x in Exercise 17 above.

```
>> y = x.^3 - 2

y =

      -1         6        25        62       123
214     341       510       727

>> plot(x,y)
```

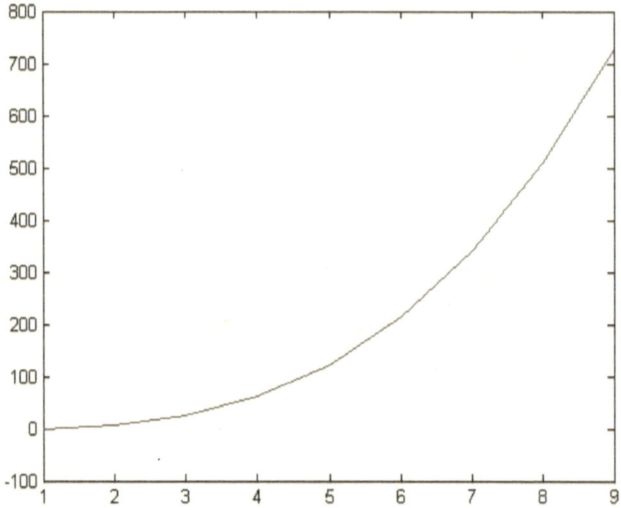

19. Generate a 4 x 4 magic square. What is the total of each row, column, and diagonal in this matrix.

```
>> magic(4)

ans =

    16     2     3    13
     5    11    10     8
     9     7     6    12
     4    14    15     1
```

The sum of each row, column, or diagonal is 34.

2. Chapter 2

1. Perform the addition operation 7+9.

```
>> 7+9

ans =

    16
```

2. Perform the subtraction operation 16-10.

```
>> 16-10

ans =

     6
```

3. Perform the multiplication operation 2*9.

```
>> 2*9

ans =

    18
```

4. Perform the division operation 12/3.

```
>> 12/3

ans =

     4
```

5. Perform the division operation 12/5.

```
>> 12/5

ans =

    2.4000
```

6. Perform the exponentiation operation 3^5.

```
>> 3^5

ans =

   243
```

7. Perform the exponentiation operation 3*(-5).

```
>> 3^(-5)

ans =

    0.0041
```

8. Perform the exponentiation operation (-3)^5.

```
>> (-3)^5

ans =

  -243
```

9. Perform the exponentiation operation -3^5.

```
>> -3^5

ans =

  -243
```

10. Compute the value of $\frac{2\pi}{3}$.

```
>> 2*pi/3

ans =
```

```
        2.0944
```

11. Obtain the value of the smallest number that can be handled by MATLAB.

```
>> eps

ans =

   2.2204e-016
```

12. Perform the multiple operations 5+7-15.

```
>> 5+7-15

ans =

    -3
```

13. Perform the multiple operations (6*7)+4.

```
>> 6*7 +4

ans =

    46
```

14. Perform the multiple operations 6*(7+4).

```
>> 6*(7+4)

ans =

    66
```

15. Perform the multiple operations 4.5 + (15/2).

```
>> 4.5 + 15/2

ans =

    12
```

16. Perform the multiple operations (4.5 + 15)/2.

```
>> (4.5 + 15)/2

ans =

    9.7500
```

17. Perform the multiple operations (15 – 4 + 12)/5 – 2*(7^4)/100.

```
>> (15-4+12)/5 - 2*(7^4)/100

ans =

  -43.4200
```

18. Perform the multiple operations (15 – 4) + 12/5 – (2*7)^4/100.

```
>> (15-4) + 12/5 - (2*7)^4/100

ans =

 -370.7600
```

19. Define the number 2/3 as a symbolic number.

```
>> sym(2/3)

ans =
```

```
2/3
```

20. Perform the fraction addition (2/3) + (3/4) numerically.

```
>> 2/3 + 3/4

ans =

        1.4167
```

21. Perform the fraction addition (2/3) + (3/4) symbolically.

```
>> sym((2/3)+(3/4))

ans =

17/12
```

3. Chapter 3

1. Perform the operation `2*3+7` and store the result in the variable `w`.

```
>> 2*3+7

ans =

    13
```

2. Define the three variables `a`, `b`, and `c` equal to `4`, `-10`, and `3.2`, respectively.

```
>> a = 4

a =
```

```
      4

>> b = -10

b =

   -10

>> c = 3.2

c =

    3.2000
```

3. Define the two variables `y` and `Y` equal to `10` and `100`. Are the two variables identical?

```
>> y = 10

y =

    10

>> Y = 100

Y =

   100
```

The two variables are not identical.

4. Let `x = 5.5` and `y = -2.6`. Calculate the value of the variable `z = 2x-3y`.

```
x =

    5.5000
```

```
>> y = -2.6

y =

   -2.6000

>> z = 2*x - 3*y

z =

   18.8000
```

5. In Exercise 4 above, calculate the value of the variable `w = 3y - z + x/y`.

```
>> w = 3*y - z + x/y

w =

  -28.7154
```

6. Let `r = 6.3` and `s = 5.8`. Calculate the value of the variable `final` defined by `final = r + s - r*s`.

```
>> r = 6.3

r =

    6.3000

>> s = 5.8

s =

    5.8000

>> final = r + s - r*s
```

```
final =

  -24.4400
```

7. In Exercise 6 above, calculate the value of the variable `this_is_the_result` defined by `this_is_the_result = r^2 - s^2`.

```
>> this_is_the_result = r^2 - s^2

this_is_the_result =

    6.0500
```

8. Define the three variable `width`, `Width`, and `WIDTH` equal to 1.5, 2.0, and 4.5, respectively. Are these three variables identical?

```
>> width = 1.5

width =

    1.5000

>> Width = 2.0

Width =

     2

>> WIDTH = 4.5

WIDTH =

    4.5000
```

The three variables are not identical.

9. Write the following comment in MATLAB: `This line will not be executed.`

```
% This line will not be executed.
```

10. Assign the value of `3.5` to the variable `s` then add a comment about this assignment on the same line.

```
>> s = 3.5      % the variable s is assigned the value 3.5

s =

    3.5000
```

11. Define the values of the variables `y1` and `y2` equal to 7 and 9 then perform the calculation `y3 = y1 - y2/3`. (Note: 2 in the formula is a subscript and should not be divided by 3).

```
>> y1 = 7

y1 =

     7

>> y2 = 9

y2 =

     9

>> y3 = y1 - y2/3
```

```
y3 =

     4
```

12. Perform the operation `2*m - 5`. Do you get an error? Why?

```
>> 2*m - 5
???    Undefined    function    or
variable 'm'.
```

We get an error message because the variable `m` is not defined.

13. Define the variables `cost` and `profit` equal to `175` and `25`, respectively, then calculate the variable `sale_price` defined by `sale_price = cost + profit.`

```
>> cost = 175

cost =

   175

>> profit = 25

profit =

    25

>> sale_price = cost + profit

sale_price =

   200
```

14. Define the variable `centigrade` equal to `28` then calculate the variable `fahrenheit` defined by `fahrenheit = (centigrade*9/5) + 32`.

```
>> centigrade = 28

centigrade =

    28

>> fahrenheit = (centigrade*9/5)
+ 32

fahrenheit =

   82.4000
```

15. Use the `format short` and `format long` commands to write the values of `14/9` to four decimals and sixteen digits, respectively.

```
>> format long
>> 14/9

ans =

   1.555555555555556

>> format short
>> 14/9

ans =

    1.5556
```

16. Perform the `who` command to get a list of the variables stored in this session.

```
>> who

Your variables are:

WIDTH                    fahrenheit
x
Width              final
y
Y                            profit
y1
a                                 r
y2
ans                               s
y3
b                        sale_price
z
c                this_is_the_result
centigrade
w                              cost
width
```

17. Perform the whos command to get a list of the variables stored in this session along with their details.

```
>> whos
  Name                Size        Bytes
Class

  WIDTH              1x1              8
double array
  Width              1x1              8
double array
  Y                   1x1             8
double array
  a                   1x1             8
double array
```

```
  ans              1x1           8
double array
  b                 1x1          8
double array
  c                 1x1          8
double array
  centigrade    1x1              8
double array
  cost             1x1           8
double array
  fahrenheit    1x1              8
double array
  final            1x1           8
double array
  profit          1x1            8
double array
  r                 1x1          8
double array
  s                 1x1          8
double array
  sale_price    1x1              8
double array
  this_is_the_result   1x1       8
double array
  w                 1x1          8
double array
  width            1x1           8
double array
  x                 1x1          8
double array
  y                 1x1          8
double array
  y1               1x1           8
double array
  y2               1x1           8
double array
  y3               1x1           8
double array
```

```
  z                        1x1              8
double array

Grand total is 24 elements using
192 bytes
```

18. Clear all the variables stored in this session by using the `clear` command.

```
>> clear
>>
```

19. Calculate both the area and perimeter of a rectangle of sides 5 and 7. No units are used in this exercise.

```
a =

       5

>> b = 7

b =

       7

>> area = a*b

area =

      35

>> perimeter = 2*(a+b)

perimeter =

      24
```

20. Calculate both the area and perimeter of a circle of radius 6.45. No units are used in this exercise.

```
>> r = 6.45

r =

     6.4500

>> area = pi*r^2

area =

   130.6981

>> perimeter = 2*pi*r

perimeter =

    40.5265
```

21. Define the symbolic variables x and z with values 4/5 and 14/17.

```
>> x = 4/5

x =

     0.8000

>> sym(x)

ans =

4/5


>> z = 14/17
```

```
z =

      0.8235

>> sym(z)

ans =

14/17
```

22. In Exercise 21 above, calculate symbolically the value of the symbolic variable y defined by y = 2x – z.

```
>> y = sym(2*x - z)

y =

66/85
```

23. Calculate symbolically the area of a circle of radius 2/3 without obtaining a numerical value. No units are used in this exercise.

```
>> radius1 = 2/3

radius1 =

      0.6667

>> radius1 = sym(radius1)

radius1 =

2/3
```

```
>> area = pi*radius1^2

area =

4/9*pi
```

24. Calculate symbolically the volume of a sphere of radius 2/3 without obtaining a numerical value. No units are used in this exercise.

```
>> radius2 = 2/3

radius2 =

    0.6667

>> radius2 = sym(radius2)

radius2 =

2/3


>> volume = (4/3)*pi*radius2^3

volume =

32/81*pi
```

25. In Exercise 23 above, use the `double` command to obtain the numerical value of the answer.

```
>> double(area)

ans =

    1.3963
```

26. In Exercise 24 above, use the `double` command to obtain the numerical value of the answer.

```
>> double(volume)

ans =

     1.2411
```

27. Define the symbolic variables `y` and `date` without assigning any numerical values to them.

```
>> y = sym('y')

y =

y


>> date = sym('date')

date =

date
```

4. Chapter 4

1. Compute the square root of 10.

```
>> sqrt(10)

ans =

     3.1623
```

2. Compute the factorial of 7.

```
>> factorial(7)
```

```
ans =

          5040
```

3. Compute the cosine of the angle 45 where 45 is in radians.

```
>> cos(45)

ans =

    0.5253
```

4. Compute the cosine of the angle 45 where 45 is in degrees.

```
>> cos(45*pi/180)

ans =

    0.7071
```

5. Compute the sine of the angle of 45 where 45 is in degrees.

```
>> sin(45*pi/180)

ans =

    0.7071
```

6. Compute the tangent of the angle 45 where 45 is in degrees.

```
>> tan(45*pi/180)

ans =
```

```
      1.0000
```

7. Compute the inverse tangent of 1.5.

```
>> atan(1.5)

ans =

      0.9828
```

The above result is in radians.

8. Compute the tangent of the angle $\frac{3\pi}{2}$. Do you get an error? Why?

```
>> tan(3*pi/2)

ans =

   5.4437e+015
```

We get a very large number approaching infinity because the tanget function is not define at $\frac{3\pi}{2}$.

9. Compute the value of exponential function e^3.

```
>> exp(3)

ans =

     20.0855
```

10. Compute the value of the natural logarithm $\ln 3.5$.

```
>> log(3.5)

ans =

    1.2528
```

11. Compute the value of the logarithm $\log_{10} 3.5$.

```
>> log10(3.5)

ans =

    0.5441
```

12. Use the MATLAB rounding function `round` to round the value of 2.43.

```
>> round(2.43)

ans =

     2
```

13. Use the MATLAB remainder function `rem` to obtain the remainder when dividing 5 by 4.

```
>> rem(5,4)

ans =

     1
```

14. Compute the absolute value of -3.6.

```
>> abs(-3.6)

ans =
```

```
    3.6000
```

15. Compute the value of the expression $1.5-2\sqrt{6.7/5}$.

```
>> 1.5 - 2*sqrt(6.7/5)

ans =

   -0.8152
```

16. Compute the value of $\sin^2\pi+\cos^2\pi$.

```
>> sin(pi)^2 + cos(pi)^2

ans =

     1
```

17. Compute the value of $\log_{10}0$. Do you get an error? Why?

```
>> log10(0)
Warning: Log of zero.
> In log10 at 17

ans =

  -Inf
```

We get minus infinity because the logarithmic function is not defined at zero.

18. Let $x = \frac{3\pi}{2}$ and $y = 2\pi$. Compute the value of the expression $2\sin x \cos y$.

```
>> x = 3*pi/2

x =

    4.7124

>> y = 2*pi

y =

    6.2832

>> 2*sin(x)*cos(y)

ans =

    -2
```

19. Compute the value of $\sqrt{45}$ symbolically and simplify the result.

```
>> sym(sqrt(45))

ans =

sqrt(45)


>> simplify(ans)

ans =

3*5^(1/2)
```

20. Compute the value of $\sqrt{45}$ numerically.

```
>> double(ans)

ans =

    6.7082
```

21. Compute the sine of the angle 45 (degrees) symbolically.

```
>> sym(sin(45*pi/180))

ans =

sqrt(1/2)
```

22. Compute the cosine of the angle 45 (degrees) symbolically.

```
> sym(cos(45*pi/180))

ans =

sqrt(1/2)
```

23. Compute the tangent of the angle 45 (degrees) symbolically.

```
>> sym(tan(45*pi/180))

ans =

1
```

24. Compute the value of $e^{\pi/2}$ symbolically.

```
>> sym(exp(pi/2))

ans =

5416116035097439*2^(-50)
```

25. Compute the value of $e^{\pi/2}$ numerically.

```
>> double(ans)

ans =

 4.8105
```

5. Chapter 5

1. Store the vector [2 4 -6 0] in the variable w.

```
>> w = [2 4 -6 0]

w =

     2     4    -6     0
```

2. In Exercise 1 above, extract the second element of the vector w.

```
> w(2)

ans =

     4
```

3. In Exercise 1 above, generate the vector z where $z = \frac{\pi}{2} w$.

```
>> z = pi*w/2

z =

     3.1416       6.2832      -9.4248
0
```

4. In Exercise 3 above, extract the fourth element of the vector z.

```
>> z(4)

ans =

        0
```

5. In Exercise 3 above, extract the first three elements of the vector z.

```
>> z(1:3)

ans =

     3.1416    6.2832   -9.4248
```

6. In Exercise 3 above, find the length of the vector z.

```
>> length(z)

ans =

        4
```

7. In Exercise 3 above, find the total sum of the values of the elements of the vector z.

```
>> sum(z)
```

```
ans =

     0
```

8. In Exercise 3 above, find the minimum and maximum values of the elements of the vector z.

```
>> min(z)

ans =

   -9.4248

>> max(z)

ans =

    6.2832
```

9. Generate a vector r with real values between 1 and 10 with an increment of 2.5.

```
>> r = (1:2.5:10)

r =

    1.0000      3.5000      6.0000
8.5000
```

10. Generate a vector s with real values of ten numbers that are equally spaced between 1 and 100.

```
>> s = linspace(1,100,10)

s =
```

```
        1        12        23        34        45
   56        67        78        89       100
```

11. Form a new vector by joining the two vectors [9 3 -2 5 0] and [1 2 -4].

```
>> a = [9 3 -2 5 0]

a =

        9         3        -2         5         0

>> b = [1 2 -4]

b =

        1         2        -4

>> c = [a b]

c =

        9         3        -2         5         0
   1        2        -4
```

12. Form a new vector by joining the vector [9 3 -2 5 0] with the number 4.

```
>> d = [a 4]

d =

        9         3        -2         5         0
   4
```

13. Add the two vectors [0.2 1.3 -3.5] and [0.5 -2.5 1.0].

```
>> x = [0.2 1.3 -3.5]

x =

    0.2000    1.3000   -3.5000

>> y = [0.5 -2.5 1.0]

y =

    0.5000   -2.5000    1.0000

>> x+y

ans =

    0.7000   -1.2000   -2.5000
```

14. Subtract the two vectors in Exercise 13 above.

```
>> x-y

ans =

   -0.3000    3.8000   -4.5000
```

15. Try to multiply the two vectors in Exercise 13 above. Do you get an error message? Why?

```
>> x*y
??? Error using ==> mtimes
Inner matrix dimensions must agree.
```

We get an error because the two vectors do not have the same length.

16. Multiply the two elements in Exercise 13 above element by element.

```
>> x.*y

ans =

    0.1000   -3.2500   -3.5000
```

17. Divide the two elements in Exercise 13 above element by element.

```
>> x./y

ans =

    0.4000   -0.5200   -3.5000
```

18. Find the dot product of the two vectors in Exercise 13 above.

```
>> x*y'

ans =

   -6.6500
```

19. Try to add the two vectors [1 3 5] and [3 6]. Do you get an error message? Why?

```
>> r = [ 1 3 5 ]

r =

     1     3     5

>> s = [ 3 6 ]
```

```
s =

     3     6

>> r+s
??? Error using ==> plus
Matrix dimensions must agree.
```

We get an error message because the two vectors do not have the same length.

20. Try to subtract the two vectors in Exercise 20 above. Do you get an error message? Why?

```
>> r-s
??? Error using ==> minus
Matrix dimensions must agree.
```

We get an error message because the two vectors do not have the same length.

21. Let the vector w be defined by w = [0.1 1.3 -2.4]. Perform the operation of scalar addition 5+w.

```
>> w = [0.1 1.3 -2.4]

w =

    0.1000    1.3000   -2.4000

>> 5+w

ans =

    5.1000    6.3000    2.6000
```

22. In Exercise 22 above, perform the operation of scalar subtraction -2-w.

```
>> 2-w

ans =

    1.9000    0.7000    4.4000
```

23. In Exercise 22 above, perform the operation of scalar multiplication 1.5*w.

```
>> 1.5*w

ans =

    0.1500    1.9500   -3.6000
```

24. In Exercise 22 above, perform the operation of scalar division w/10.

```
>> w/10

ans =

    0.0100    0.1300   -0.2400
```

25. In Exercise 22 above, perform the operation 3 – 2*w/5.

```
>> 3-2*w/5

ans =

    2.9600    2.4800    3.9600
```

26. Define the vector b by b = [0 pi/3 2pi/3 pi]. Evaluate the three vectors $\sin b$, $\cos b$, and $\tan b$ (element by element).

```
>> b = [0 pi/3 2*pi/3 pi]

b =

         0    1.0472    2.0944
3.1416

>> sin(b)

ans =

         0    0.8660    0.8660
0.0000

>> cos(b)

ans =

    1.0000    0.5000   -0.5000
-1.0000

>> tan(b)

ans =

         0    1.7321   -1.7321
-0.0000
```

27. In Exercise 26 above, evaluate the vector e^b (element by element).

```
>> exp(b)

ans =

    1.0000    2.8497    8.1205
23.1407
```

28. In Exercise 26 above, evaluate the vector $\sqrt{b}$ (element by element).

```
>> sqrt(b)

ans =

         0    1.0233    1.4472
1.7725
```

29. Try to evaluate the vector 3^b. Do you get an error message? Why?

```
>> 3^b
??? Error using ==> mpower
Matrix must be square.
```

Yes, because this operation needs to be performed element by element.

30. Perform the operation in Exercise 29 above element by element?

```
>> 3.^b

ans =

    1.0000    3.1597    9.9834
31.5443
```

31. Generate a vector of 1's with a length of 4 elements.

```
> ones(1,4)

ans =

     1     1     1     1
```

32. Generate a vector of 0's with a length of 6 elements.

```
>> zeros(1,6)

ans =

       0       0       0       0       0
0
```

33. Sort the elements of the vector [0.35 -1.0 0.24 1.30 -0.03] in ascending order.

```
>> k = [0.35 -1.0 0.24 1.30 -
0.03]

k =

    0.3500    -1.0000    0.2400
1.3000   -0.0300

>> sort(k)

ans =

   -1.0000    -0.0300    0.2400
0.3500    1.3000
```

34. Generate a random permutation vector with 5 elements.

```
>> randperm(5)

ans =

     2     4     3     5     1
```

35. For the vector [2 4 -3 0 1 5 7], determine the range, mean, and median.

```
>> v = [2 4 -3 0 1 5 7]

v =

        2        4       -3        0        1
5       7

>> range(v)

ans =

    10

>> mean(v)

ans =

    2.2857

>> median(v)

ans =

     2
```

36. Define the symbolic vector x = [r s t u v].

```
>> syms r s t u v
>> x = [r s t u v]

x = [ r, s, t, u, v]
```

37. In Exercise 36 above,. perform the addition operation of the two vectors x and [1 0 -2 3 5] to obtain the new symbolic vector y.

```
>> r = [1 0 -2 3 5]

r =

      1      0     -2      3      5

>> y = x+r

y =

[ r+1,   s, t-2, u+3, v+5]
```

38. In Exercise 37 above, extract the third element of the vector y.

```
>> y(3)

ans =

t-2
```

39. In Exercise 37 above, perform the operation 2*x/7 + 3*y.

```
>> 2*x/7 + 3*y

ans =

[  23/7*r+3,      23/7*s,   23/7*t-
6,  23/7*u+9, 23/7*v+15]
```

40. In Exercise 37 above, perform the dot products x*y' and y*x'. Are the two results the same? Why?

```
>> x*y'
```

```
ans =

r*(1+conj(r))+s*conj(s)+t*(-
2+conj(t))+u*(3+conj(u))+v*(5+co
nj(v))


>> y*x'

ans =

(r+1)*conj(r)+s*conj(s)+(t-
2)*conj(t)+(u+3)*conj(u)+(v+5)*c
onj(v)
```

The two results are different because MATLAB treats these symbolic variables as complex variables by default. If they were real variables, then the results would be the same.

41. In Exercise 37 above, find the square root of the symbolic vector x+y.

```
>> sqrt(x+y)

ans =

[   (2*r+1)^(1/2), 2^(1/2)*s^(1/2),
(2*t-2)^(1/2),    (2*u+3)^(1/2),
(2*v+5)^(1/2)]
```

6. Chapter 6

1. >> x = [25 18 17 24 11 10 20 30 25 12 11 12 15 28 33 20 12 20 22 30 15 18 22 19 31]

```
x =

  Columns 1 through 11

    25      18      17      24      11
10    20    30    25    12    11

  Columns 12 through 22

    12      15      28      33      20
12    20    22    30    15    18

  Columns 23 through 25

    22    19    31
```

```
   2. >> size(x)

ans =

     1    25
```

```
   3. >> mean(x)

ans =

    20

>> median(x)

ans =

    20

>> mode(x)

ans =
```

```
12
```

9. >> range(x)

```
ans =

    23

>> iqr(x)

ans =

   10.7500

>> var(x)

ans =

   47.9167

>> std(x)

ans =

    6.9222
```

9. >> max(x)

```
ans =

   33

>> min(x)

ans =

   10
```

9. >> quantile(x,0.30)

```
ans =

    15

>> quantile(x,0.50)

ans =

    20

>> quantile(x,0.70)

ans =

    24
```

9.

```
>> prctile(x,47)

ans =

   19.2500

>> prctile(x,55)

ans =

    20

>> prctile(x,80)

ans =

   26.5000
```

9.

```
>> zscore(x)

ans =

  Columns 1 through 7
```

```
   0.7223    -0.2889    -0.4334
0.5779      -1.3002     -1.4446
0

  Columns 8 through 14

   1.4446     0.7223    -1.1557
-1.3002      -1.1557     -0.7223
1.1557

  Columns 15 through 21

   1.8780          0    -1.1557
0    0.2889    1.4446   -0.7223

  Columns 22 through 25

  -0.2889     0.2889    -0.1445
1.5891
```

```
9.          >>  x  =  [1.24  2.56
            3.45  4.66  3.56  4.90
            3.33  2.45  2.34  2.00
            1.45 1.56]

x =

  Columns 1 through 7

   1.2400     2.5600     3.4500
4.6600       3.5600      4.9000
3.3300

  Columns 8 through 12

   2.4500     2.3400     2.0000
1.4500    1.5600

>> y = [600 450 320 300 430 330
470 570 340 200 250 390]
```

```
y =

  Columns 1 through 11

   600      450      320      300      430
330      470      570      340      200
250

  Column 12

   390

>> corrcoef(x,y)

ans =

    1.0000    -0.1906
   -0.1906     1.0000
```

Thus the correlation coefficient in this Exercise turns out to be -0.1906 indicating a weak negative correlation. In the next Exercise, the same correlation coefficient is calculated from scratch, but this time using the formula of Pearson's correlation coefficient (see Chapter 12).

10. Step One: Enter the data points given in the Exercise:

```
>> x = [1.24 2.56 3.45 4.66 3.56
4.90 3.33 2.45 2.34 2.00 1.45 1.56]

x =

  Columns 1 through 7

    1.2400          2.5600          3.4500
4.6600          3.5600          4.9000
3.3300
```

```
  Columns 8 through 12

    2.4500         2.3400         2.0000
1.4500    1.5600

>> y = [600 450 320 300 430 330 470
570 340 200 250 390]

y =

  Columns 1 through 11

   600      450      320      300      430
330    470    570    340    200    250

  Column 12

   390
```

Step Two: Calculate the five different sums needed in the formula for the correlation coefficient:

```
>> sumx = sum(x)

sumx =

   33.5000

>> sumy = sum(y)

sumy =

        4650

>> mxx = x.*x

mxx =

  Columns 1 through 7
```

```
    1.5376         6.5536        11.9025
21.7156       12.6736          24.0100
11.0889

  Columns 8 through 12

    6.0025        5.4756          4.0000
2.1025    2.4336

>> sumxx = sum(mxx)

sumxx =

  109.4960

>> myy = y.*y

myy =

  Columns 1 through 5

      360000                         202500
102400        90000        184900

  Columns 6 through 10

      108900                         220900
324900       115600        40000

  Columns 11 through 12

       62500       152100

>> sumyy = sum(myy)

sumyy =

     1964700

>> mxy = x.*y

mxy =
```

```
   1.0e+03 *

  Columns 1 through 7

    0.7440       1.1520       1.1040
1.3980       1.5308       1.6170
1.5651

  Columns 8 through 12

    1.3965       0.7956       0.4000
0.3625    0.6084

>> sumxy = sum(mxy)

sumxy =

   1.2674e+04

>> sumx2 = sumx*sumx

sumx2 =

   1.1223e+03

>> sumy2 = sumy*sumy

sumy2 =

    21622500

>> n = size(x)

n =

     1    12

>> n1 = n(2)

n1 =
```

```
12
```

Step Three: Use the formula to compute the value of the correlation coefficient based on the sums obtained above:

```
>> r = (n1*sumxy - sumx*sumy) / (sqrt(n1*sumxx-
sumx2)*sqrt(n1*sumyy-sumy2))

r =

   -0.1906
```

It is clear from the above result that the correlation between the two data sets is negative and weak. The value of the correlation coefficient is close to 0 indicating a weak correlation. It is the same value obtained in Exercise 9 above that was obtained directly using the MATLAB command `corrcoef`.

11. The same value of the correlation coefficient is obtained in both Exercises 9 and 10 above. The first value in Exercise 9 was obtained directly using the MATLAB command `corrcoef`. The second value in Exercise 10 was obtained from scratch using the formula for Pearson's correlation coefficient.

7. Chapter 7

1.
```
>> x = [25 18 17 24 11 10 20 30 25
12 11 12 15 28 33 20 12 20 22 30 15
18 22 19 31]

x =

  Columns 1 through 11
```

```
    25        18        17        24        11        10
20      30      25      12      11

  Columns 12 through 22

    12        15        28        33        20        12
20      22      30      15      18

  Columns 23 through 25

    22      19      31

>> bar(x)
>> hold on
>> xlabel('Data')
>> ylabel('Number')
```

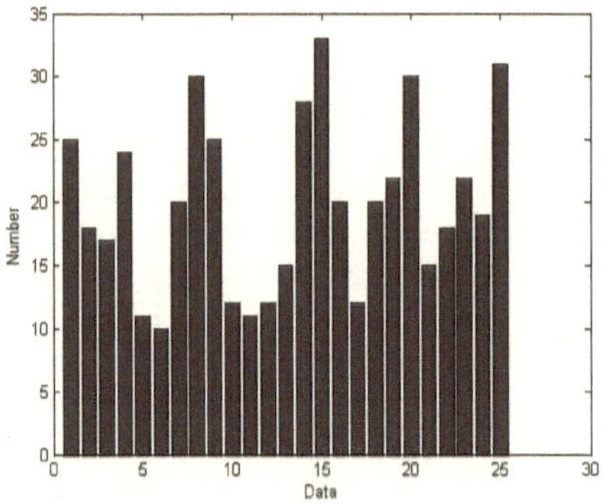


```
   2. >> hist(x)
>> hold on
>> xlabel('Data')
>> ylabel('Number')
```

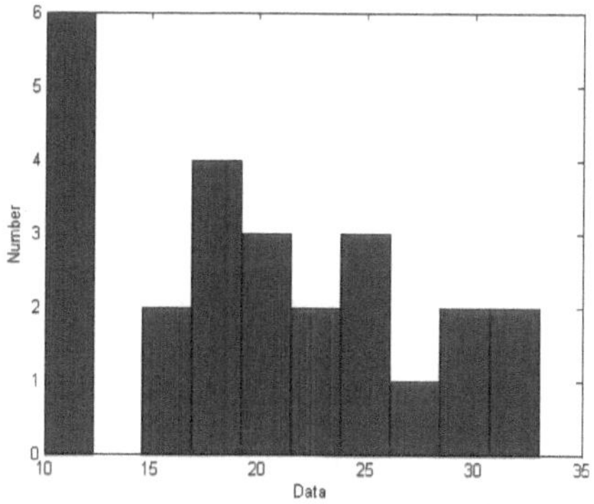

3. >> cdfplot(x)

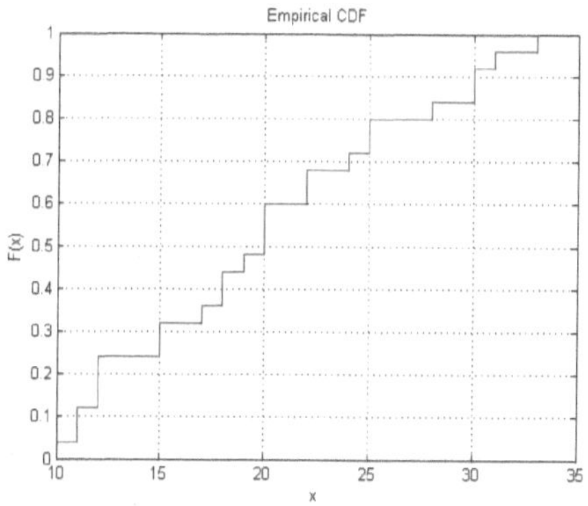

4. >> size(x)

```
ans =

     1    25

>> y = 1:25

y =
```

```
  Columns 1 through 11

     1         2         3         4         5         6
7      8       9      10      11

  Columns 12 through 22

    12        13        14        15        16        17
18      19      20      21      22

  Columns 23 through 25

    23      24      25

>> scatter(y,x)
>> hold on
>> xlabel('Data')
>> ylabel('Number')
```

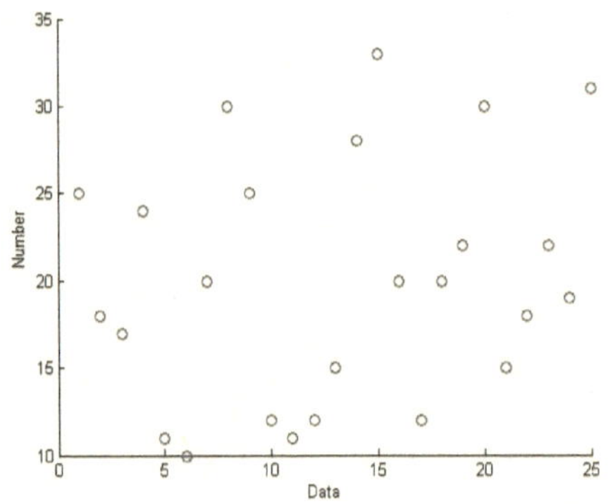

5. >> pie(x)

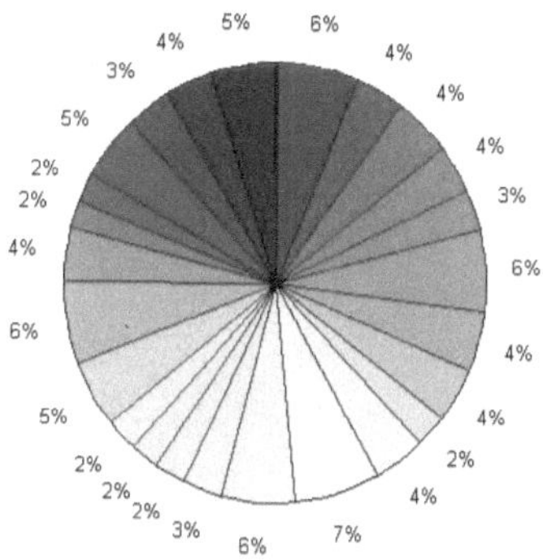

6. >> rose(x)

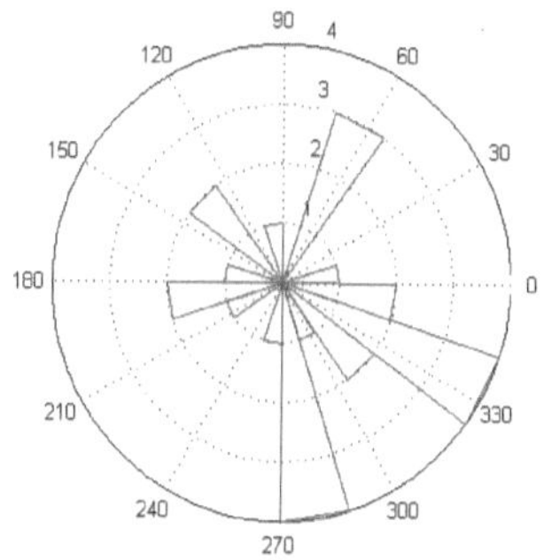

7. >> stem(x)

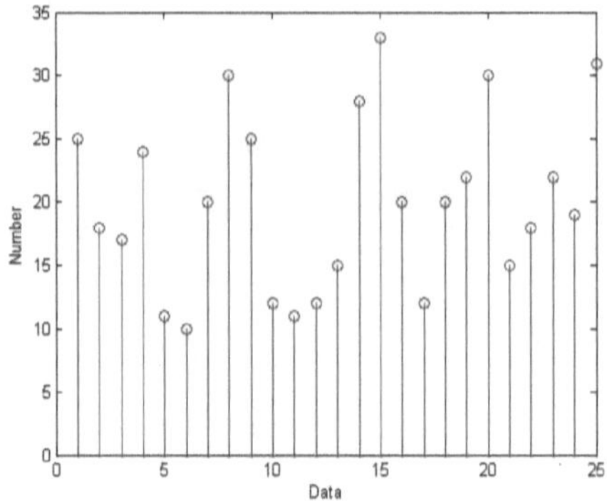

8. >> zscore(x)

```
ans =

  Columns 1 through 7

    0.7223   -0.2889   -0.4334    0.5779
-1.3002   -1.4446         0

  Columns 8 through 14

    1.4446    0.7223   -1.1557   -1.3002
-1.1557   -0.7223    1.1557

  Columns 15 through 21

    1.8780         0   -1.1557         0
0.2889    1.4446   -0.7223

  Columns 22 through 25

   -0.2889    0.2889   -0.1445    1.5891

>> scatter(y,ans)
>> hold on
>> xlabel('Data')
>> ylabel('zscores')
```

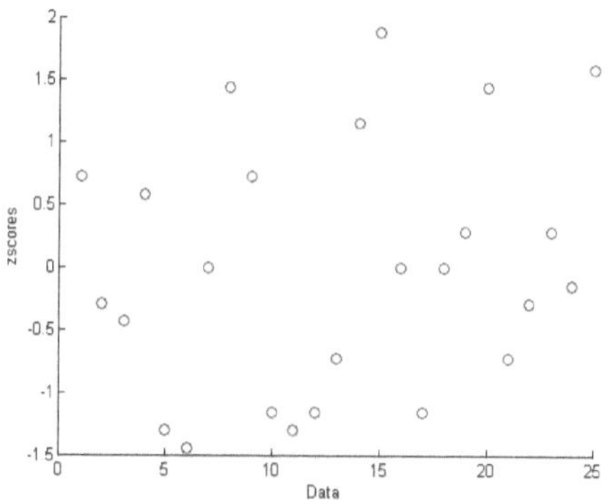


```
9. >> boxplot(x)
>> hold on
>> ylabel('Data')
```

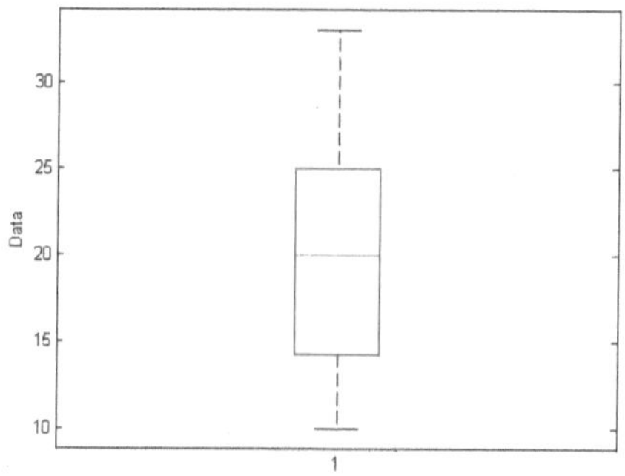

8. Chapter 8

1. Step One: Enter the given data points and calculate the different required sums:

```
>> x = [50 52 56 61 66 70 80 84 88 92]
```

```
x =

    50      52      56      61      66      70
80     84     88     92

>> y = [200 220 250 310 360 400 420 440
500 550]

y =

   200     220     250     310     360     400
420    440    500    550

>> sumx = sum(x)

sumx =

   699

>> sumy = sum(y)

sumy =

          3650

>> mxx = x.*x

mxx =

  Columns 1 through 6

          2500                2704                3136
3721            4356            4900

  Columns 7 through 10

          6400                7056                7744
8464

>> sumxx = sum(mxx)

sumxx =
```

```
       50981

>> mxy = x.*y

mxy =

  Columns 1 through 5

       10000           11440           14000
18910        23760

  Columns 6 through 10

       28000           33600           36960
44000        50600

>> sumxy = sum(mxy)

sumxy =

      271270

>> sumx2 = sumx*sumx

sumx2 =

      488601

>> n = size(x)

n =

     1    10

>> n1 = n(2)

n1 =

    10
```

Step Two: Calculate the two constants *a* and *b* using the two formulas given:

```
>> a = (sumxx*sumy - sumx*sumxy)/(n1*sumxx - sumx2)

a =

 -166.7726

>> b = (n1*sumxy - sumx*sumy)/(n1*sumxx - sumx2)

b =

    7.6076
```

Step Three: Write the equation of the best fitting line as follows:

$$y = -166.7726 + 7.6076x$$

Step Four: Plot the best fitting line along with the give data points on the same graph for comparison:

```
>> y1 = a + b*x

y1 =

  Columns 1 through 7

  213.6084   228.8236   259.2541   297.2922   335.3303   365.7608   441.8370

  Columns 8 through 10

  472.2674   502.6979   533.1284

>> plot(x,y1)
>> hold on
```

```
>> scatter(x,y)
>> xlabel('x')
>> ylabel('y')
```

The resulting graph is shown in Figure 14.2.

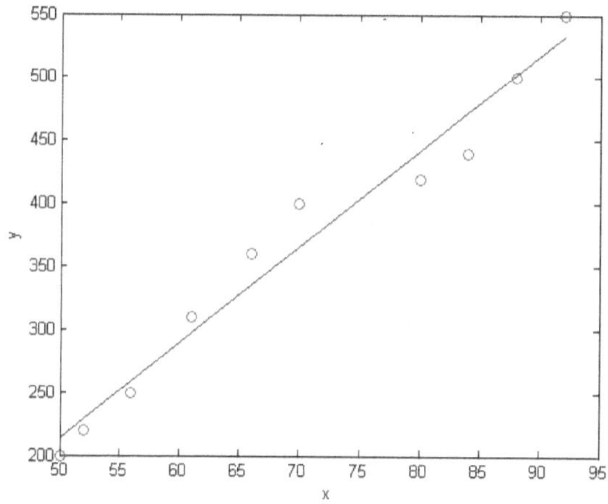

2. Step One: Enter the given data points and calculate the different sums needed:

```
>> x = [10 20 25 31 36 44 55 60 62 65 70
75]

x =

    10      20      25      31      36      44
55      60      62      65      70      75

>> y = [15 25 30 35 40 50 60 65 70 75 80
85]

y =

    15      25      30      35      40      50
60      65      70      75      80      85
```

```
>> sumx = sum(x)

sumx =

   553

>> sumy = sum(y)

sumy =

   630

>> mxx = x.*x

mxx =

  Columns 1 through 6

          100          400          625
961          1296          1936

  Columns 7 through 12

         3025          3600          3844
4225          4900          5625

>> sumxx = sum(mxx)

sumxx =

       30537

>> mxy = x.*y

mxy =

  Columns 1 through 5

          150          500          750
1085          1440
```

```
  Columns 6 through 10

        2200            3300            3900
4340        4875

  Columns 11 through 12

        5600        6375

>> sumxy = sum(mxy)

sumxy =

       34515

>> sumx2 = sumx*sumx

sumx2 =

      305809

>> n = size(x)

n =

     1    12

>> n1 = n(2)

n1 =

    12
```

Step Two: Calculate the two constants *a* and *b* as follows:

```
>>        a        =        (sumxx*sumy        -
sumx*sumxy)/(n1*sumxx - sumx2)

a =

    2.4988
```

```
>> b = (n1*sumxy - sumx*sumy)/(n1*sumxx -
sumx2)

b =

    1.0850
```

Step 3: Write the equation of the best fitting line as follows:

$$y = 2.4988 + 1.0850x$$

Step 4: Plot the best fitting line along with the given date points on the same graph for comparison:

```
>> y1 = a + b*x

y1 =

  Columns 1 through 7

   13.3490    24.1991    29.6242    36.1343
41.5594   50.2395   62.1747

  Columns 8 through 12

   67.5998    69.7699    73.0249    78.4500
83.8751

>> plot(x,y1)
>> hold on
>> scatter(x,y)
>> xlabel('x')
>> ylabel('y')
```

The resulting graph is shown in Figure 14.3.

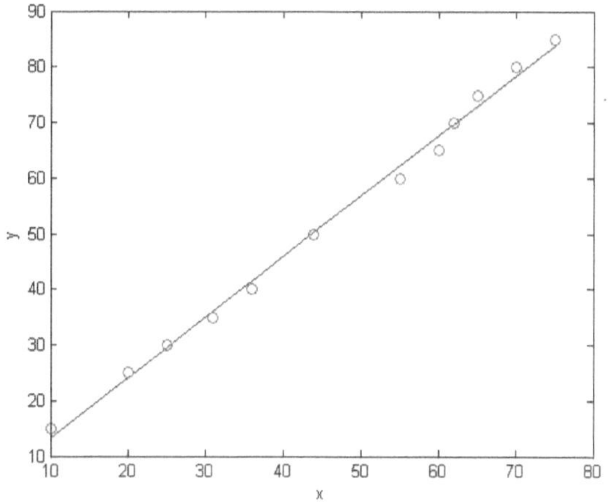

3.. Let us start with the quadratic equation given in the Exercise as follows:

$$y = a + bx + cx^2$$

where a, b, and c are constant coefficients to be determined for a specific set of data. We will proceed with the detailed derivation of the equations of simple nonlinear regression.

For a set of data x_i and y_i (where $i = 1,2,3,......n$), the above equation becomes for each point (x_i, y_i):

$$y_i = a + bx_i + cx_i^2$$

(AA)

It should be noted that the above equations is repeated n times, where n is the number of data points. Let us write the above equations for the first few data points as $i = 1,2,3,......n$:

$$y_1 = a + bx_1 + cx_1^2$$

$$y_2 = a + bx_2 + cx_2^2$$

$$y_3 = a + bx_3 + cx_3^2$$

....

....

....

$$y_n = a + bx_n + cx_n^2$$

Let us now add the above list of equations to obtain the following expression:

$$y_1 + y_2 + y_3 + ... + y_n = a + bx_1 + cx_1^2 + a + bx_2 + cx_2^2 + a + bx_3 + cx_3^2 + ... + a + bx_n + cx_n^2$$

The above equation can be written in the following simplified form:

$$y_1 + y_2 + y_3 + ... + y_n = (a + a + a + ... + a) + (bx_1 + bx_2 + bx_3 + ... + bx_n) + (cx_1^2 + cx_2^2 + cx_3^2 + ... + cx$$

where the sum of the a's in the first parenthesis is n times. The above equation can be written using the summation sign $\sum$ as follows:

$$\sum_{i=1}^{n} y_i = na + b\sum_{i=1}^{n} x_i + c\sum_{i=1}^{n} x_i^2$$
(BB)

where the following summation notation is used in equation (BB):

$$\sum_{i=1}^{n} y_i = y_1 + y_2 + y_3 + \ldots\ldots.. + y_n$$

$$na = a + a + a + \ldots.. + a \qquad (summed\ \ n\ \ times)$$

$$b\sum_{i=1}^{n} x_i = b\left(x_1 + x_2 + x_3 + \ldots\ldots + x_n\right) = bx_1 + bx_2 + bx_3 + \ldots. + bx_n$$

$$c\sum_{i=1}^{n} x_i^2 = c\left(x_1^2 + x_2^2 + x_3^2 + \ldots.. + x_n^2\right) = cx_1^2 + cx_2^2 + cx_3^2 + \ldots.. + cx_n^2$$

The next step would be to multiply equation (AA) by x_i to obtain the following equation:

$$x_i y_i = ax_i + bx_i x_i + cx_i^2 x_i = ax_i + bx_i^2 + cx_i^3$$
(CC)

Writing the above equation out for each value $i = 1, 2, 3, ...n$, we obtain:

$$x_1 y_1 = a x_1 + b x_1^2 + c x_1^3$$

$$x_2 y_2 = a x_2 + b x_2^2 + c x_2^3$$

$$x_3 y_3 = a x_3 + b x_3^2 + c x_3^3$$

....

....

....

$$x_n y_n = a x_n + b x_n^2 + c x_n^3$$

Adding the above list of equations results in the following expression:

$$x_1 y_1 + x_2 y_2 + x_3 y_3 + ... + x_n y_n = a x_1 + b x_1^2 + c x_1^3 + a x_2 + b x_2^2 + \\ + a x_3 + b x_3^2 + c x_3^3 + ... + a x_n + b x_n^2 + c x_n^3$$

The above equation can be simplified as follows:

$$x_1 y_1 + x_2 y_2 + x_3 y_3 + ... + x_n y_n = a x_1 + a x_2 + a x_3 + ... + a x_n \\ + b x_1^2 + b x_2^2 + b x_3^2 + ... + b x_n^2 + c x_1^3 + c x_2^3 + c x_3^3 + ...$$

The above equation can be written using the summation sign $\sum$ as follows:

$$\sum_{i=1}^{n} x_i y_i = a\sum_{i=1}^{n} x_i + b\sum_{i=1}^{n} x_i^2 + c\sum_{i=1}^{n} x_i^3$$
(DD)

where the following summation notation is used in equation (DD):

$$\sum_{i=1}^{n} x_i y_i = x_1 y_1 + x_2 y_2 + x_3 y_3 + \ldots\ldots + x_n y_n$$

$$a\sum_{i=1}^{n} x_i = a\left(x_1 + x_2 + x_3 + \ldots\ldots + x_n\right) = ax_1 + ax_2 + ax_3 + \ldots\ldots + ax_n$$

$$b\sum_{i=1}^{n} x_i^2 = b\left(x_1^2 + x_2^2 + x_3^2 + \ldots\ldots + x_n^2\right) = bx_1^2 + bx_2^2 + bx_3^2 + \ldots\ldots + bx_n^2$$

$$c\sum_{i=1}^{n} x_i^3 = c\left(x_1^3 + x_2^3 + x_3^3 + \ldots\ldots + x_n^3\right) = cx_1^3 + cx_2^3 + cx_3^3 + \ldots\ldots + cx_n^3$$

The next step would be to multiply equation (AA) by x_i^2 to obtain the following equation:

$$x_i^2 y_i = ax_i^2 + bx_i x_i^2 + cx_i^2 x_i^2 = ax_i^2 + bx_i^3 + cx_i^4$$
(EE)

Writing the above equation out for each value $i = 1, 2, 3, \ldots\ldots n$, we obtain:

$$x_1^2 y_1 = a x_1^2 + b x_1^3 + c x_1^4$$

$$x_2^2 y_2 = a x_2^2 + b x_2^3 + c x_2^4$$

$$x_3^2 y_3 = a x_3^2 + b x_3^3 + c x_3^4$$

....

....

....

$$x_n^2 y_n = a x_n^2 + b x_n^3 + c x_n^4$$

Adding the above list of equations results in the following expression:

$$x_1^2 y_1 + x_2^2 y_2 + x_3^2 y_3 + \ldots.. + x_n^2 y_n = a x_1^2 + b x_1^3 + c x_1^4 + a x_2^2 + b x_2^3 + \\ + a x_3^2 + b x_3^3 + c x_3^4 + \ldots\ldots\ldots + a x_n^2 + b x_n^3 + c x_n^4$$

The above equation can be simplified as follows:

$$x_1^2 y_1 + x_2^2 y_2 + x_3^2 y_3 + \ldots.. + x_n^2 y_n = a x_1^2 + a x_2^2 + a x_3^2 + \ldots\ldots + a x_n^2 \\ + b x_1^3 + b x_2^3 + b x_3^3 + \ldots\ldots + b x_n^3 + c x_1^4 + c x_2^4 + c x_3^4 + \ldots\ldots +$$

The above equation can be written using the summation sign $\sum$ as follows:

$$\sum_{i=1}^{n} x_i^2 y_i = a\sum_{i=1}^{n} x_i^2 + b\sum_{i=1}^{n} x_i^3 + c\sum_{i=1}^{n} x_i^4$$
(FF)

where the following summation notation is used in equation (FF):

$$\sum_{i=1}^{n} x_i^2 y_i = x_1^2 y_1 + x_2^2 y_2 + x_3^2 y_3 + \ldots\ldots + x_n^2 y_n$$

$$a\sum_{i=1}^{n} x_i^2 = a\left(x_1^2 + x_2^2 + x_3^2 + \ldots\ldots + x_n^2\right) = ax_1^2 + ax_2^2 + ax_3^2 + \ldots\ldots + ax_n^2$$

$$b\sum_{i=1}^{n} x_i^3 = b\left(x_1^3 + x_2^3 + x_3^3 + \ldots\ldots + x_n^3\right) = bx_1^3 + bx_2^3 + bx_3^3 + \ldots\ldots + bx_n^3$$

$$c\sum_{i=1}^{n} x_i^4 = c\left(x_1^4 + x_2^4 + x_3^4 + \ldots\ldots + x_n^4\right) = cx_1^4 + cx_2^4 + cx_3^4 + \ldots\ldots + cx_n^4$$

Next we solve equations (BB), (DD), and (FF) simultaneously. The three equations are summarized below. The three equations can then be solve simultaneously for the three constants a, b, and c. However, this is not performed here as it is not required by the Exercise. Usually, the various sums are computed first, the inserted into the three equations directly. This is followed by solving the three equations numerically using the MATLAB command `solve` as explained in Chapter 10 on Solving Equations.

$$\sum_{i=1}^{n} y_i = na + b\sum_{i=1}^{n} x_i + c\sum_{i=1}^{n} x_i^2$$

$$\sum_{i=1}^{n} x_i y_i = a\sum_{i=1}^{n} x_i + b\sum_{i=1}^{n} x_i^2 + c\sum_{i=1}^{n} x_i^3$$

$$\sum_{i=1}^{n} x_i^2 y_i = a\sum_{i=1}^{n} x_i^2 + b\sum_{i=1}^{n} x_i^3 + c\sum_{i=1}^{n} x_i^4$$

9. Chapter 9

1. Step One: Enter the given data points and calculate the eight different sums needed as follows:

   1.1: Enter the given data points:

```
>> x = [109 118 124 100 90 120 150]

x =

   109      118      124      100        90
120    150

>>  y  =  [450  440  500  600  450  550
600]

y =

   450      440      500      600      450
550    600

>> z = [11 16 17 20 22 26 31]

z =

    11       16       17       20       22
26     31
```

   1.2: Calculate the eight different sums needed for this exercise:

```
>> sumx = sum(x)

sumx =

   811
```

```
>> sumy = sum(y)

sumy =

        3590

>> sumz = sum(z)

sumz =

   143

>> mxx = x.*x

mxx =

  Columns 1 through 5

       11881                        13924
15376        10000          8100

  Columns 6 through 7

       14400       22500

>> sumxx = sum(mxx)

sumxx =

       96181

>> mxy = x.*y

mxy =

  Columns 1 through 5

       49050                        51920
62000        60000         40500

  Columns 6 through 7
```

```
        66000        90000

>> sumxy = sum(mxy)

sumxy =

      419470

>> mxz = x.*z

mxz =

  Columns 1 through 5

        1199                              1888
2108          2000          1980

  Columns 6 through 7

        3120         4650

>> sumxz = sum(mxz)

sumxz =

       16945

>> myz = y.*z

myz =

  Columns 1 through 5

        4950                              7040
8500         12000          9900

  Columns 6 through 7

       14300        18600

>> sumyz = sum(myz)
```

```
sumyz =

        75290

>> myy = y.*y

myy =

  Columns 1 through 5

      202500                               193600
250000        360000        202500

  Columns 6 through 7

      302500        360000

>> sumyy = sum(myy)

sumyy =

     1871100

>> n = size(x)

n =

     1     7

>> n1 = n(2)

n1 =

     7
```

Step Two: Write the system of three simultaneous governing equations for the three constants a, b, and c as follows:

$$7a + 811b + 3590c = 143$$

$$811a + 96181b + 419470c = 16945$$

$$3590a + 419470b + 1891100c = 75290$$

Step Three: Solve the system of three simultaneous algebraic equations and obtain the values of the three constants a, b, and c as follows:

```
>> A = [n1 sumx sumy ; sumx sumxx
sumxy ; sumy sumxy sumyy]

A =

           7                      811
3590
         811                    96181
419470
        3590                   419470
1871100

>> b = [sumz ; sumxz ; sumyz]

b =

         143
       16945
       75290

>> inv(A)*b

ans =

  -17.4828
    0.0813
    0.0555

>> a = ans(1)

a =
```

```
   -17.4828

>> b = ans(2)

b =

     0.0813

>> c = ans(3)

c =

     0.0555
```

Step Four: Write the equation of the best fitting plane as follows:

$$z = -17.4828 + 0.0813x + 0.0555y$$

No need to plot a graph of the plane or estimate the error in the analysis. But we can calculate the values of z using the best fitting plane equation as follows:

```
>> z1 = a + b*x + c*y

z1 =

   16.3792         16.5558         20.3767
23.9795         14.8339         22.8287
28.0462
```

The approximate calculated values shown above are reasonable and are very close to the data values given in the exercise.

2. <u>Step One:</u> Let us enter the data points given as well as calculate the twelve different sums needed:

2.1: Let us first enter the data points given in the Exercise:

```
>> x = [234 245 257 270 250 270
299]

x =

   234     245     257     270     250
270   299

>> y = [2 4 7 9 7 11 13]

y =

     2       4       7       9       7
11     13

>> z = [890 810 910 900 1100 1200
1000]

z =

  Columns 1 through 6

           890                        810
910                 900               1100
1200

  Column 7

          1000

>> w = [1030 1100 1200 1350 1440
1500· 1570]

w =

  Columns 1 through 6
```

```
        1030                                  1100
1200                    1350                  1440
1500

  Column 7

        1570
```

2.2: Let us calculate the twelve different sums needed in this Exercise:

```
>> sumx = sum(x)

sumx =

        1825

>> sumy = sum(y)

sumy =

    53

>> sumz = sum(z)

sumz =

        6810

>> sumw = sum(w)

sumw =

        9190

>> mxx = x.*x

mxx =

  Columns 1 through 5
```

```
       54756                      60025
66049        72900        62500

  Columns 6 through 7

       72900        89401

>> sumxx = sum(mxx)

sumxx =

      478531

>> myy = y.*y

myy =

     4      16      49      81      49
121   169

>> sumyy = sum(myy)

sumyy =

   489

>> mzz = z.^z

mzz =

   Inf     Inf     Inf     Inf     Inf
Inf   Inf

>> mzz = z.*z

mzz =

  Columns 1 through 5

      792100                     656100
828100       810000      1210000
```

```
  Columns 6 through 7

     1440000     1000000

>> sumzz = sum(mzz)

sumzz =

     6736300

>> mxy = x.*y

mxy =

  Columns 1 through 5

         468                             980
1799          2430          1750

  Columns 6 through 7

        2970          3887

>> sumxy = sum(mxy)

sumxy =

       14284

>> mxz = x.*z

mxz =

  Columns 1 through 5

      208260                          198450
233870        243000        275000

  Columns 6 through 7

      324000        299000
```

```
>> sumxz = sum(mxz)

sumxz =

     1781580

>> myz = y.*z

myz =

  Columns 1 through 5

        1780                         3240
6370         8100          7700

  Columns 6 through 7

       13200        13000

>> sumyz = sum(myz)

sumyz =

       53390

>> mxw = x.*w

mxw =

  Columns 1 through 5

      241020                       269500
308400       364500       360000

  Columns 6 through 7

      405000       469430

>> sumxw = sum(mxw)

sumxw =
```

```
    2417850

>> myw = y.*w

myw =

  Columns 1 through 5

       2060                                  4400
8400        12150          10080

  Columns 6 through 7

      16500        20410

>> sumyw = sum(myw)

sumyw =

      74000

>> mzw = z.*w

mzw =

  Columns 1 through 5

     916700                              891000
1092000      1215000      1584000

  Columns 6 through 7

    1800000      1570000

>> sumzw = sum(mzw)

sumzw =

    9068700

>> n =size(x)
```

```
n =

     1     7

>> n1 = n(2)

n1 =

     7
```

Step Two: Let us write the system of four simultaneous algebraic equations for the four constants a, b, c, and d as follows:

$$7a + 1825b + 53c + 6810d = 9190$$

$$1825a + 478531b + 14284c + 1781580d = 2417850$$

$$53a + 14284b + 489c + 53390d = 74000$$

$$6810a + 1781580b + 53390c + 6736200d = 9068700$$

Step Three: Let us solve the above system of four simultaneous algebraic equations for the four constants a, b, c, and d using the inverse matrix method (check Chapter 10 on Solving Equations for details):

```
>> A = [n1 sumx sumy sumz ; sumx
sumxx sumxy sumxz ; sumy sumxy
sumyy sumyz ; sumz sumxz sumyz
sumzz]

A =
```

```
           7          1825            53
6810
        1825        478531         14284
1781580
          53         14284           489
53390
        6810       1781580         53390
6736300

>> b = [sumw ; sumxw ; sumyw ;
sumzw]

b =

        9190
     2417850
       74000
     9068700

>> inv(A)*b

ans =

 -121.7772
    2.4977
   24.2467
    0.6166

>> a = ans(1)

a =

 -121.7772

>> b = ans(2)

b =

     2.4977

>> c = ans(3)

c =
```

```
    24.2467

>> d = ans(4)

d =

     0.6166
```

Step Four: Let us write the equation of the best fitting hyperplane as follows:

$$w = -121.7772 + 2.4977x + 24.2467y + 0.6166z$$

Step Five: Let us calcuale the approximate values for w to obtained from the best fitting hyperplane equation above:

```
>> w1 = a + b*x + c*y + d*z

w1 =

    1.0e+03 *

     1.0600          1.0866          1.2510
1.3258          1.3506          1.5592
1.5568
```

It is seen from the above results that the approximate values obtained through the best fit are very close to the actual data points given in the Exercise. In this case, there is no need to generate a graph because MATLAB does not have a `plot4` command. Also, there is no need to calculate the errors.

3. Consider a set of n data points (x_i, y_i, z_i, w_i) in four dimensions. In this case and since we are dealing with four-dimensional space, we obtain a best fitting hyperplane instead of a best fitting line. The equation of the best fitting hyperplane can be written in general as follows:

$$w = a + bx + cy + dz$$

What is required is find the values of the four constants a, b, c. and d The same procedure of Chapter 15 is followed here but note that we are dealing with four dimensions instead of three dimensions. The above equation is repeated below with subscript $i = 1,2,3,......n$:

$$w_i = a + bx_i + cy_i + dz_i$$

(E*)

It should be noted that the above equations is repeated n times, where n is the number of data points. Let us write the above equations for the first few data points as $i = 1,2,3,......n$:

$$w_1 = a + bx_1 + cy_1 + dz_1$$

$$w_2 = a + bx_2 + cy_2 + dz_2$$

$$w_3 = a + bx_3 + cy_3 + dz_3$$

....

....

....

$$w_n = a + b\,x_n + c\,y_n + d\,z_n$$

Let us now add the above list of equations to obtain the following expression:

$$w_1 + w_2 + w_3 + \ldots\ldots + w_n = a + b\,x_1 + c\,y_1 + d\,z_1 + a + b\,x_2 + c\,y_2 + d\,z_2 + a + b\,x_3 + c\,y_3 + d\,z_3 \ldots\ldots + a + b\,x_n + c\,y_n + d\,z_n$$

The above equation can be written in the following simplified form:

$$w_1 + w_2 + w_3 + \ldots\ldots + w_n = (a + a + a + \ldots. + a) + (b\,x_1 + b\,x_2 + b\,x_3 + \ldots\ldots + b\,x_n) + (c\,y_1 + c\,y_2 + c\,y_3 + \ldots\ldots + c\,y_n) + (d\,z_1 + d\,z_2 + d\,z_3 + \ldots\ldots + d\,z_n)$$

where the sum of the a's in the first parenthesis is n times. The above equation can be written using the summation sign $\sum$ as follows:

$$\sum_{i=1}^{n} w_i = na + b\sum_{i=1}^{n} x_i + c\sum_{i=1}^{n} y_i + d\sum_{i=1}^{n} z_i$$

(F*)

where the following summation notation is used in equation (F*):

$$\sum_{i=1}^{n} w_i = w_1 + w_2 + w_3 + \dots\dots + w_n$$

$$na = a + a + a + \dots\dots + a \qquad (summed\ \ n\ \ times)$$

$$b\sum_{i=1}^{n} x_i = b\left(x_1 + x_2 + x_3 + \dots\dots + x_n\right) = b\,x_1 + b\,x_2 + b\,x_3 + \dots + b$$

$$c\sum_{i=1}^{n} y_i = c\left(y_1 + y_2 + y_3 + \dots\dots + y_n\right) = c\,y_1 + c\,y_2 + c\,y_3 + \dots + c$$

$$d\sum_{i=1}^{n} z_i = d\left(z_1 + z_2 + z_3 + \dots\dots + z_n\right) = d\,z_1 + d\,z_2 + d\,z_3 + \dots + d$$

The next step would be to multiply equation (E*) by x_i to obtain the following equation:

$$x_i\,w_i = a\,x_i + b\,x_i\,x_i + c\,x_i\,y_i + d\,x_i\,z_i$$

(G*)

Writing the above equation out for each value $i = 1, 2, 3, \dots\dots n$, we obtain:

$$x_1\,w_1 = a\,x_1 + b\,x_1^2 + c\,x_1\,y_1 + d\,x_1\,z_1$$

$$x_2\,w_2 = a\,x_2 + b\,x_2^2 + c\,x_2\,y_2 + d\,x_2\,z_2$$

$$x_3 w_3 = a x_3 + b x_3^2 + c x_3 y_3 + d x_3 z_3$$

....

....

....

$$x_n w_n = a x_n + b x_n^2 + c x_n y_n + d x_n z_n$$

Adding the above list of equations results in the following expression:

$$\begin{aligned} x_1 w_1 + x_2 w_2 + x_3 w_3 + \ldots\ldots + x_n w_n &= a x_1 + b x_1^2 + c x_1 y_1 + d x_1 z_1 \\ &+ a x_2 + b x_2^2 + c x_2 y_2 + d x_2 z_2 + a x_3 + b x_3^2 + c x_3 y_3 + d x_3 z_3 \\ &+ \ldots\ldots + a x_n + b x_n^2 + c x_n y_n + d x_n z_n \end{aligned}$$

The above equation can be simplified as follows:

$$\begin{aligned} x_1 w_1 + x_2 w_2 + x_3 w_3 + \ldots\ldots + x_n w_n &= a x_1 + a x_2 + a x_3 + \ldots\ldots + a x_n \\ &+ b x_1^2 + b x_2^2 + b x_3^2 + \ldots\ldots + b x_n^2 \\ &+ c x_1 y_1 + c x_2 y_2 + c x_3 y_3 + \ldots\ldots + c x_n y_n \\ &+ d x_1 z_1 + d x_2 z_2 + d x_3 z_3 + \ldots\ldots + d x_n z_n \end{aligned}$$

The above equation can be written using the summation sign $\sum$ as follows:

$$\sum_{i=1}^{n} x_i\, w_i = a\sum_{i=1}^{n} x_i + b\sum_{i=1}^{n} x_i^2 + c\sum_{i=1}^{n} x_i\, y_i + d\sum_{i=1}^{n} x_i\, z_i$$
(H*)

where the following summation notation is used in equation (H*):

$$\sum_{i=1}^{n} x_i\, w_i = x_1\, w_1 + x_2\, w_2 + x_3\, w_3 + \ldots\ldots + x_n w_n$$

$$a\sum_{i=1}^{n} x_i = a\left(x_1 + x_2 + x_3 + \ldots\ldots + x_n\right) = a\,x_1 + a\,x_2 + a\,x_3 + \ldots\ldots +$$

$$b\sum_{i=1}^{n} x_i^2 = b\left(x_1^2 + x_2^2 + x_3^2 + \ldots\ldots + x_n^2\right) = b\,x_1^2 + b\,x_2^2 + b\,x_3^2 + \ldots\ldots + b$$

$$c\sum_{i=1}^{n} x_i\, y_i = c\left(x_1\, y_1 + x_2\, y_2 + x_3\, y_3 + \ldots\ldots + x_n\, y_n\right)$$
$$= c\,x_1\, y_1 + c\,x_2\, y_2 + c\,x_3\, y_3 + \ldots\ldots + c\,x_n\, y_n$$

$$d\sum_{i=1}^{n} x_i\, z_i = d\left(x_1\, z_1 + x_2\, z_2 + x_3\, z_3 + \ldots\ldots + x_n\, z_n\right)$$
$$= d\,x_1\, z_1 + d\,x_2\, z_2 + d\,x_3\, z_3 + \ldots\ldots + d\,x_n\, z_n$$

The next step would be to multiply equation (E*) by y_i to obtain the following equation:

$$y_i\, w_i = a\,y_i + b\,x_i\, y_i + c\,y_i\, y_i + d\,y_i\, z_i$$
(J*)

Writing the above equation out for each value $i = 1, 2, 3, \ldots\ldots n$, we obtain:

$$y_1 w_1 = a\, y_1 + b\, x_1\, y_1 + c\, y_1^2 + d\, y_1\, z_1$$

$$y_2 w_2 = a\, y_2 + b\, x_2\, y_2 + c\, y_2^2 + d\, y_2\, z_2$$

$$y_3 w_3 = a\, y_3 + b\, x_3\, y_3 + c\, y_3^2 + d\, y_3\, z_3$$

$$\ldots\ldots\ldots\ldots\ldots\ldots$$

$$\ldots\ldots\ldots\ldots\ldots\ldots$$

$$\ldots\ldots\ldots\ldots\ldots\ldots$$

$$y_n w_n = a\, y_n + b\, x_n\, y_n + c\, y_n^2 + d\, y_n\, z_n$$

Adding the above list of equations results in the following expression:

$$\begin{aligned} y_1 w_1 + y_2 w_2 + y_3 w_3 + \ldots\ldots + y_n w_n &= a\, y_1 + b\, x_1\, y_1 + c\, y_1^2 + d\, y_1\, z_1 \\ &+ a\, y_2 + b\, x_2\, y_2 + c\, y_2^2 + d\, y_2\, z_2 + a\, x_3 + b\, x_3\, y_3 + c\, y_3^2 + d\, y_3\, z_3 \\ &+ \ldots\ldots + a\, y_n + b\, x_n\, y_n + c\, y_n^2 + d\, y_n\, z_n \end{aligned}$$

The above equation can be simplified as follows:

$$
\begin{aligned}
y_1 w_1 + y_2 w_2 + y_3 w_3 + \ldots\ldots + y_n w_n &= a y_1 + a y_2 + a y_3 + \ldots\ldots + a y_n \\
&+ b x_1 y_1 + b x_2 y_2 + b x_3 y_3 + \ldots\ldots + b x_n y_n \\
&+ c y_1^2 + c y_2^2 + c y_3^2 + \ldots\ldots + c y_n^2 \\
&+ d y_1 z_1 + d y_2 z_2 + d y_3 z_3 + \ldots\ldots + d y_n z_n
\end{aligned}
$$

The above equation can be written using the summation sign $\sum$ follows:

$$\sum_{i=1}^{n} y_i w_i = a\sum_{i=1}^{n} y_i + b\sum_{i=1}^{n} x_i y_i + c\sum_{i=1}^{n} y_i^2 + d\sum_{i=1}^{n} y_i z_i$$

(K*)

where the following summation notation is used in equation (H):

$$\sum_{i=1}^{n} y_i w_i = y_1 w_1 + y_2 w_2 + y_3 w_3 + \ldots\ldots + y_n w_n$$

$$a\sum_{i=1}^{n} y_i = a(y_1 + y_2 + y_3 + \ldots\ldots + y_n) = a y_1 + a y_2 + a y_3 + \ldots\ldots + a y_n$$

$$
\begin{aligned}
b\sum_{i=1}^{n} x_i y_i &= b(x_1 y_1 + x_2 y_2 + x_3 y_3 + \ldots\ldots + x_n y_n) \\
&= b x_1 y_1 + b x_2 y_2 + b x_3 y_3 + \ldots\ldots + b x_n y_n
\end{aligned}
$$

$$c\sum_{i=1}^{n} y_i^2 = c(y_1^2 + y_2^2 + y_3^2 + \ldots\ldots + y_n^2) = c y_1^2 + c y_2^2 + c y_3^2 + \ldots\ldots + c y_n^2$$

$$d\sum_{i=1}^{n} y_i z_i = d(y_1 z_1 + y_2 z_2 + y_3 z_3 + ... + y_n z_n)$$

$$= d y_1 z_1 + d y_2 z_2 + d y_3 z_3 + ... + d y_n z_n$$

The next step would be to multiply equation (E*) by z_i to obtain the following equation:

$$z_i w_i = a z_i + b x_i z_i + c y_i z_i + d z_i^2$$
(L*)

Writing the above equation out for each value $i = 1, 2, 3, ... n$, we obtain:

$$z_1 w_1 = a z_1 + b x_1 z_1 + c y_1 z_1 + d z_1^2$$

$$z_2 w_2 = a z_2 + b x_2 z_2 + c y_2 z_2 + d z_2^2$$

$$z_3 w_3 = a z_3 + b x_3 z_3 + c y_3 z_3 + d z_3^2$$

....

....

....

$$z_n w_n = a z_n + b x_n z_n + c y_n z_n + d z_n^2$$

Adding the above list of equations results in the following expression:

$$z_1 w_1 + z_2 w_2 + z_3 w_3 + \ldots\ldots + z_n w_n = a z_1 + b x_1 z_1 + d y_1 z_1 + d z_1^2$$
$$+ a z_2 + b x_2 z_2 + c y_2 z_2 + d z_2^2 + a z_3 + b x_3 z_3 + c y_3 z_3 + d z$$
$$+ \ldots\ldots + a z_n + b x_n z_n + c y_n z_n + d z_n^2$$

The above equation can be simplified as follows:

$$z_1 w_1 + z_2 w_2 + z_3 w_3 + \ldots\ldots + z_n w_n = a z_1 + a z_2 + a z_3 + \ldots\ldots + a z$$
$$+ b x_1 z_1 + b x_2 z_2 + b x_3 z_3 + \ldots\ldots + b x_n z_n$$
$$+ c y_1 z_1 + c y_2 z_2 + c y_3 z_3 + \ldots\ldots + c y_n z_n$$
$$+ d z_1^2 + d z_2^2 + d z_3^2 + \ldots\ldots + d z_n^2$$

The above equation can be written using the summation sign $\sum$ as follows:

$$\sum_{i=1}^{n} z_i w_i = a \sum_{i=1}^{n} z_i + b \sum_{i=1}^{n} x_i z_i + c \sum_{i=1}^{n} y_i z_i + d \sum_{i=1}^{n} z_i^2$$

(M*)

where the following summation notation is used in equation (M*):

$$\sum_{i=1}^{n} z_i w_i = z_1 w_1 + z_2 w_2 + z_3 w_3 + \ldots\ldots + z_n w_n$$

$$a \sum_{i=1}^{n} z_i = a(z_1 + z_2 + z_3 + \ldots\ldots + z_n) = a z_1 + a z_2 + a z_3 + \ldots\ldots + a z_n$$

$$b\sum_{i=1}^{n} x_i\, z_i = b\left(x_1\, z_1 + x_2\, z_2 + x_3\, z_3 + ... + x_n\, z_n\right)$$

$$= b\, x_1\, z_1 + b\, x_2\, z_2 + b\, x_3\, z_3 + ... + b\, x_n\, z_n$$

$$c\sum_{i=1}^{n} y_i\, z_i = c\left(y_1\, z_1 + y_2\, z_2 + y_3\, z_3 + ... + y_n\, z_n\right)$$

$$= c\, y_1\, z_1 + c\, y_2\, z_2 + c\, y_3\, z_3 + ... + c\, y_n\, z_n$$

$$d\sum_{i=1}^{n} z_i^2 = d\left(z_1^2 + z_2^2 + z_3^2 + ... + z_n^2\right) = d\, z_1^2 + d\, z_2^2 + d\, z_3^2 + ... + d\, z_n^2$$

Next we list equations (F*), (H*), (K*), and (M*) as shown in the boxed list below. These four equations should be solved simultaneously to obtain the four constants a, b, c, and d.

$$\sum_{i=1}^{n} w_i = na + b\sum_{i=1}^{n} x_i + c\sum_{i=1}^{n} y_i + d\sum_{i=1}^{n} z_i$$
(F*)

$$\sum_{i=1}^{n} x_i w_i = a\sum_{i=1}^{n} x_i + b\sum_{i=1}^{n} x_i^2 + c\sum_{i=1}^{n} x_i y_i + d\sum_{i=1}^{n} x_i z_i$$
(H*)

$$\sum_{i=1}^{n} y_i w_i = a\sum_{i=1}^{n} y_i + b\sum_{i=1}^{n} x_i y_i + c\sum_{i=1}^{n} y_i^2 + d\sum_{i=1}^{n} y_i z_i$$
(K*)

$$\sum_{i=1}^{n} z_i w_i = a\sum_{i=1}^{n} z_i + b\sum_{i=1}^{n} x_i z_i + c\sum_{i=1}^{n} y_i z_i + d\sum_{i=1}^{n} z_i^2$$
(M*)

References

MATLAB Tutorials

1. http://www.cyclismo.org/tutorial/matlab/

   This is an online tutorial in HTML format from Union College, New York. The tutorial emphasizes vectors, matrices, vector operations, loops, plots, executable files (scripts), subroutines (functions), if statements, and data files.

2. http://www.math.ufl.edu/help/matlab-tutorial/

   This is an online tutorial in HTML format from the University of Florida, Gainesville. The tutorial emphasizes matrices, variables, functions, decisions, loops, and scripts. This tutorial has a nice list of important commands of MATLAB with a short description of each command.

3. http://www.mines.utah.edu/gg_computer_seminar/matlab/matlab.html

   This is an online tutorial in HTML format available at the website of University of Utah, Salt Lake City (Original by Kermit Sigmon, University of Florida). This tutorial emphasizes matrices, decisions, loops, scalar functions, vector functions, matrix functions, strings, and graphics. You may also download the tutorial as a postscript file (39 pages that include a comprehensive reference section).

4. http://web.mit.edu/afs/.athena/astaff/project/logos/olh/Math/Matlab/Matlab.html

   This is an online tutorial in HTML format hosted at MIT. It emphasizes matrices, arithmetic and logical operators, control structures, selective indexing, polynomial operations, signal processing functions, graphics, scripts, and functions.

5. http://www.mit.edu/people/abbe/matlab/main.html

   This is an online 3-day tutorial at MIT in HTML format. It covers matrices, various MATLAB commands, help, plotting, polynomials, variables, scripts, functions, loops, debugging, differential equations, vectorization, three-dimensional graphics, and symbolic math.

6. http://texas.math.ttu.edu/~gilliam/ttu/m4330/m4330.html

   This is a tutorial hosted at Texas Tech University, Lubbock. They provide six PDF files for download in the form of six lessons for a mathematics course they teach there. These lessons are somewhat advanced and specialized and are not recommended for beginner students of MATLAB.

7. http://www.engin.umich.edu/group/ctm/basic/basic.html

   This is a basic tutorial from Carnegie Mellon and the University of Michigan. It covers the basics of vectors, functions, plotting, polynomials, matrices, printing, help, and M-files.

8. http://www.mathworks.com/academia/student_center/tutorials/launchpad.html

   This is an online tutorial from the MathWorks, the company that develops and sells MATLAB. This tutorial covers variables, calculations, plotting, scripts, and files.

9. http://www.math.utah.edu/lab/ms/matlab/matlab.html#starting

   This is an online short tutorial from the University of Utah, Salt Lake City. This is a very simple tutorial recommended for beginners. It covers matrices, vectors, systems of equations, loops, and graphing in one and two dimensions.

10. http://www.math.mtu.edu/~msgocken/intro/intro.html

    This is an online tutorial in HTML format hosted at Michigan Tech and written by Mark

S. Gockenback. It covers calculations, graphs, programming, advanced matrix calculations, advanced graphics, solving nonlinear problems, and advanced data types like structures, cell arrays, and objects.

11. http://www.eece.maine.edu/mm/matweb.html

    MATLAB Educational Websites: This is an online HTML page containing links to several educational websites on various topics using MATLAB. The page contains also links to numerous MATLAB tutorials.

12. http://web.cecs.pdx.edu/~gerry/MATLAB/

    MATLAB Hypertext Reference: This is an online tutorial in HTML format written by Gerald Recktenwald and hosted at Portland State University, Portland. This tutorial covers variables, plotting, and programming.

13. http://www.facstaff.bucknell.edu/maneval/help211/helpmain.html

    Helpful Information for Using MATLAB: These pages of helpful information are maintained by Jim Maneval and hosted at Bucknell University, Lewisburg.

14. http://www.indiana.edu/~statmath/math/matlab/gettingstarted/index.html

Getting Started with MATLAB: This is an online tutorial in HTML format. The tutorial may also be downloaded and printed as a PDF file (14 pages). This tutorial covers syntax, matrices, graphics, and programming.

15. www.math.udel.edu/~driscoll/teaching/**matlab**_adv.pdf

    Crash Course in MATLAB: This tutorial is in the form of a PDF file download. It is written by Tobin A. Driscoll and hosted at the University of Delaware, The tutorial is 66 pages that can be downloaded and printed. This tutorial covers arrays, matrices, scripts, functions, errors, 2D and 3D graphics, color, handles and properties, vectorization, advanced data structures (strings, cell arrays, structures), linear algebra, optimization, data fitting, quadrature, and differential equations.

16. www.maths.dundee.ac.uk/~ftp/na-reports/**MatlabNotes**.pdf

    An Introduction to MATLAB: This tutorial is in the form of a PDF file download. It is written by David F. Griffiths and is hosted at the University of Dundee. The tutorial is 37 pages that can be downloaded and printed. This tutorial covers numbers, formats, variables, output, vectors, 2D plotting, scripts, vector operations, matrices, matrix operations, systems of linear equations, strings, loops, decisions, 3D plotting, files, and graphical user interfaces.

17. www.geosci.uchicago.edu/~gidon/geosci236/organize/**matlab**Intro.pdf

    An Introduction to MATLAB: This tutorial is in the form of a PDF file download. It is written by S. Butenko, P. Pardalos, and L. Pitsoulis. It is hosted at the University of Chicago. The tutorial is 28 pages that can be downloaded and printed. This tutorial covers matrices, matrix operations, workspace, formats, functions, vector functions, matrix functions, polynomial functions, programming, loops, decisions, M-files, and 2D and 3D graphics.

18. http://www.math.colostate.edu/~gerhard/classes/340/notes/index.html

    MATLAB Notes: This tutorial is in the form HTML pages online. It is hosted at Colorado State University. This tutorial covers arithmetic, array operations, scripts, functions, plotting, polynomials, matrices, vectors, linear systems of algebraic equations, and symbolic computation. There are some links to more tutorials on the internet that are hosted at other places.

19. http://www.indiana.edu/~statmath/math/matlab/gettingstarted/index.html

    Getting Started with MATLAB: This tutorial is in the form of online HTML pages that are hosted at Indiana University. It may also be downloaded as a PDF file (14 pages) and printed. This tutorial covers syntax, matrices, graphics, and programming.

20. http://www-ccs.ucsd.edu/matlab/toolbox/symbolic/symbmath.html

    This is an online reference guide for the MATLAB Symbolic Math Toolbox. It is hosted at the University of California, San Diego. The tutorial covers about 85 MATLAB symbolic math commands that are listed alphabetically with examples.

21. http://www.phys.ufl.edu/docs/matlab/toolbox/symbolic/ch1.html

    Using the Symbolic Math Toolbox: This tutorial is in the form of HTML pages that is hosted at the University of Florida, Gainesville. This tutorial seems to be comprehensive and covers calculus, simplifications, substitutions, variable-precision arithmetic, linear algebra, solving equations, special mathematical functions, using Maple functions, and the extended symbolic math toolbox.

22. http://en.wikipedia.org/wiki/Matlab

    This is a small overview in the form of an HTML page hosted at the website of Wikipedia, the free internet encyclopedia. The page has a brief history of MATLAB and some useful links.

23. http://www.facstaff.bucknell.edu/maneval/help211/exercises.html

    MATLAB Exercises: This web page provides some MATLAB exercises for practice. The site is hosted at Bucknell University. The exercises cover syntax, arrays, relational operators, logical operators, decisions, loops, and programming.

24. http://www.math.chalmers.se/~nilss/enm/INTRO_matlab_I.html

    MATLAB Exercises: This web page provides three sets of MATLAB exercises for practice. The site is hosted at Chalmers University of Technology and Gteborg University. The exercises cover arithmetic, formats, variables, vectors, matrices, decisions, scripts, loops, computer arithmetic, strings, and graphics.

25. http://college.hmco.com/mathematics/larson/elementary_linear/5e/students/matlabs.html

    MATLAB Exercises: These are MATLAB exercises taken from the book "Elementary Linear Algebra", fifth edition, by Ron Larson. Each set of exercises is downloaded as a PDF file that can be printed. These exercises cover systems of linear equations, matrices, determinants, vector spaces, inner product spaces, linear transformations, eigenvalues and eigenvectors, and complex vector spaces.

26. http://www.math.umn.edu/~ROberts/math5385/matlabEx1.html

    Supplementary MATLAB Exercises: These exercises are hosted at the University of Minnesota. They are written specifically for a mathematics course at the university. These exercises mostly deal with the graphics capabilities of MATLAB.

27. www.kom.aau.dk/~borre/**matlab**7/exercise.pdf

    This is a PDF file download of a set of 8 pages of MATLAB exercises. The exercises are somewhat advanced and are not recommended for beginning students of MATLAB.

28. www.chaos.swarthmore.edu/courses/Phys50L_2006/**Matlab/MatlabExercises**.pdf

    This is a PDF file download of 27 pages of MATLAB exercises. These exercises are written for a physics course and are somewhat advanced. They are not recommended for beginners.

MATLAB Books:

29. Gilat, A., *MATLAB: An Introduction with Applications*, Second Edition, John Wiley & Sons, 2004.

30. Pratap, R., *Getting Started with MATLAB 7: An Introduction for Scientists and Engineers*, Oxford University Press, 2005.

31. Hanselmann, D. and Littlefield, B., *Mastering MATLAB 7*, Prentice Hall, 2004.

32. Palm, W., *Introduction to MATLAB 7 for Engineers*, McGraw-Hill, 2004.

33. Moore, H., *MATLAB for Engineers*, Prentice Hall, 2006.

34. Chapman, S., *MATLAB Programming for Engineers*, Thomson Engineering, 2004.

35. Davis, T. and Sigmon, K., *MATLAB Primer*, Seventh Edition, Chapman & Hall, 2004.

36. Higham, D. and Higham, N., *MATLAB Guide*, Second Edition, SIAM, 2005.

37. King, J., *MATLAB for Engineers*, Addison-Wesley, 1988.

38. Etter, D., *Introduction to MATLAB for Engineers and Scientists*, Prentice Hall, 1995.

39. Magrab, E. et al., *An Engineer's Guide to MATLAB*, Prentice Hall, 2000.

40. Etter, D., Kuncicky, D. and Hull, D., *Introduction to MATLAB 6*, Prentice Hall, 2001.

41. Recktenwald, G., *Introduction to Numerical Methods and MATLAB: Implementation and Applications*, Prentice Hall, 2001.

42. Biran, A. and Breiner, M., *MATLAB 5 for Engineers*, Addison-Wesley, 1999.

43. Part-Enander, E. and Sjoberg, A., *The MATLAB 5 Handbook*, Addison-Wesley, 1999.

44. Etter, D., *Engineering Problem Solving with MATLAB*, Prentice Hall, 1993.

45. Chen, K., Giblin, P, and Irving, A., *Mathematical Explorations with MATLAB*, Cambridge University Press, 1999.

46. Mathews, J. and Fink, K., *Numerical Methods Using MATLAB*, Third Edition, Prentice Hall, 1999.

47. Fausett, L., *Applied Numerical Analysis Using MATLAB*, Prentice Hall, 1999.

Statistics Books:

48. Harper, W. M., *Statistics*, Fifth Edition, Longman Group UK, 1988.

49. Daniel, W. W. and Terrell, J. C., *Business Statistics: Basic Concepts and Methodology*, Third Edition, Houghton Mifflin Company, 1983.

50. Spiegel, M. R., *Theory and Problems of Statistics: Schaum's Outline Series*, McGraw-Hill, 1972.

51. Swinscow, T. D. V., *Statistics at Square One*, The British Medical Journal, 1991.

52. Freedman, D., Pisani, R., Purves, R., and Adhikari, A., *Statistics*, Second Edition, W. W. Norton & Company, 1991.

53. Wesley Barnes, J., *Statistical Analysis for Engineers and Scientists:: A Computer-Based Approach*, McGraw-Hill, 1994.

54. Bhattacharyya, G. K. and Johnson, R. A., *Statistical Concepts and Methods*, Wiley, 1077.

55. Steel, R. G. D. and Torrie, J. H., *Principles and Procedures of Statistics: A Biometric Approach*, McGraw-Hill, 1980.

56. Rumsey, D. J., *Statistics for Dummies*, Second Edition, For Dummies Series, 2011.

57. Urdan, T. C., *Statistics in Plain English*, Third Edition, Routledge, 2010.

58. White, R. S. and White, J. S., *Statistics*, Tenth Edition, Wiley, 2013.

59. McClave, J. T. and Sincich, T. T., *Statistics*, Twelfth Edition, Pearson, 2012.

60. Triola, M. F., *Elementary Statistics*, Twelfth Edition, Pearson, 2012.

Statistics/MATLAB Books:

61. Matinez, W. L. and Marinez, A. R., *Computational Statistics Handbook with MATLAB*, Third Edition, Chapman & Hall - CRC, 2015.

62. Cho, M. and Martinez, W. L., *Statistics in MATLAB: A Primer*, Chapman & Hall – CRC, 2014.

63. Lopez, C. P., *Statistics with MATLAB: Examples and Exercises*, Createspace Independent Publishing Platform, 2013.

64. Martinez, W. L. and Martinez, A. R., *Exploratory Data Analysis with MATLAB*, Second Edition, Chapman & Hall – CRC, 2010.

65. Braselton, J., *Data Analysis with MATLAB*, Createspace Independent Publishing Platform, 2016.

Regression Analysis Books:

66. Schroeder, L. D., Sjoquist, D. L., and Stepham P. E., *Understanding Regression Analysis*, SAGE Publications, 1986.

67. Montgomery, D. C. and Peck, E. A., *Introduction to Linear Regression Analysis*, Fifth Edition, Wiley, 2012.

68. Mendenhall, W. and Sincich, T. T., *A Second Course in Statistics: Regression Analysis*, Seventh Edition, Pearson, 2011.

69. Draper, N. R. and Smith, H., *Applied Regression Analysis*, Wiley-Interscience, 1998.

70. Chatterjee, S. and Hadi, A. S., *Regression Analysis by Example*, Fifth Edition, Wiley, 2012.

71. Sen, A. and Srivastava, M., *Regression Analysis: Theory, Methods, and Applications*, Springer, 1997.

72. Hoffmann, J. P. and Shafer, K., *Linear Regression Analysis: Assumptions and Applications*, NASW Press, 2015.

Regression Analysis/MATLAB Books:

73. Braselton, J., *Curve Fitting with MATLAB: Linear and Nonlinear Regression, Interpolation*, Createspace Independent Publishing Platform, 2016.

74. Matinez, W. L. and Marinez, A. R., *Computational Statistics Handbook with MATLAB*, Third Edition, Chapman & Hall - CRC, 2015.

Data Science Books

1. **Joel Grus – *Data Science from Scratch: First Principles with Python*** (2nd Ed., 2019), O'Reilly Media — A practical introduction to core algorithms and techniques implemented from first principles.
2. **Carl Shan, William Chen, Haijiang Henry Wang & Max Song – *The Data Science Handbook: Advice and Insights from 25 Amazing Data Scientists*** (2015), Data Science Bookshelf — Career perspectives and insights from experienced data science practitioners.
3. **Foster Provost & Tom Fawcett – *Data Science for Business: What You Need to Know about Data Mining and Data-Analytic Thinking*** (1st Ed., 2013), O'Reilly

Media — A foundational text connecting data science methods to business decision-making.

4. **Tony Hey, Kristin Michele Tolle & Stewart Tansley (eds.) –** ***The Fourth Paradigm: Data-Intensive Scientific Discovery*** (2009), Microsoft Research — An influential anthology on the paradigm shift toward data-intensive science.
5. **Catherine D'Ignazio & Lauren F. Klein –** ***Data Feminism*** (2020), MIT Press — Explores data science through a social justice and equity lens, addressing bias and power in data.
6. **Chris Wiggins & Matthew L. Jones –** ***How Data Happened: A History from the Age of Reason to the Age of Algorithms*** (2023), W. W. Norton & Company — A historical and societal overview of data's rise and influence.
7. **Ivo D. Dinov –** ***Data Science and Predictive Analytics: Biomedical and Health Applications using R*** (1st Ed., 2018), Springer — A textbook focusing on data science methods and predictive modeling with real applications.

Additional widely referenced books (not fully detailed here but often recommended):

- *Python for Data Analysis* by Wes McKinney (data manipulation with Python)

- *Hands-On Machine Learning with Scikit-Learn, Keras & TensorFlow* by Aurélien Géron (applied machine learning)
- *Practical Statistics for Data Scientists* (Bruce, Bruce & Gedeck) (statistics in data science)
- *Storytelling with Data* by Cole Nussbaumer Knaflic (data visualization)

Installation of MATLAB

In this book, it is assumed that you have already installed MATLAB on your computer system. To go through the eleven chapters of the book, you need to have MATLAB running on your computer. For help on installing MATLAB on your computer, check the following web links:

MATLAB Installation for Windows:

http://www.itc.virginia.edu/research/matlab/install/installwin.html

MATLAB Installation for Linux:

http://www.freebsd.org/doc/zh_TW/books/handbook/linuxemu-matlab.html

MATLAB Installation for Macintosh:

http://www.itc.virginia.edu/research/matlab/install/installmac.html

MATLAB Installation for Other Systems:

http://shum.huji.ac.il/cc/matlR131inuxhomeinst.html

Index

www.ingramcontent.com/pod-product-compliance
Lightning Source LLC
LaVergne TN
LVHW090314210926
848480LV00001B/240
* 9 7 9 8 2 9 5 6 2 1 8 2 6 *